Splicing Wire and Fiber Rope

SAILMAKER'S EYE SPLICE

(For instructions on making this splice, turn to page 54.)

SPLICING WIRE AND FIBER ROPE

Raoul Graumont and John Hensel

CORNELL MARITIME PRESS

Centreville, Maryland

ISBN 0-87033-118-3
Library of Congress Catalog Card Number: 45-3379

Manufactured in the United States of America
1994 reprint

Dedication

To former shipmates of the Cape Horn sailing days,
with whom we shared many memorable experiences in
their lifelong battle with the elements, we gratefully dedi-
cate this book.

R. G.
J. H.

Acknowledgments

In the preparation of this volume, it has been our good fortune to have received the helpful co-operation of numerous friends, yachtsmen, riggers and U.S. Army and Navy Officials. We wish to extend our gratitude and thanks to all of them.

We have also received additional aid from numerous articles in magazines and books that we wish to acknowledge.

We are likewise indebted to the Paulsen Cable Co. for the use of their cable and equipment and desire to thank them for their assistance in this respect.

For their generous assistance in supplying the Manila rope used in constructing the examples on rope splicing we wish to thank the Whitlock Cordage Co.

We extend our thanks to Mr. Newton Alfred for his preparation of the art work.

In particular, we wish to acknowledge with thanks the able assistance of Mr. Don Selchow in the preparation of the photographic work and for his generous and helpful co-operation with numerous other details.

We also thank Mr. Harry Hensel for his share of the photographs.

RAOUL GRAUMONT
JOHN HENSEL

Preface

In the long ago days of sailing ships, wire rope was almost unknown, as fiber rope was used for all types of running rigging. With the advent of steam propulsion, wire came into prominent use on all kinds of standing rigging on board steamships; as a result, wire rope today has almost entirely replaced fiber rope and chains which were formerly used for hoisting and hauling.

Sailing yachts have also found ample use for small flexible wire cable in connection with tail splicing for main halyards, backstays, jib-sheets and standing rigging.

Wire rope of every type is, in fact, becoming so widely used nowadays, that nearly every factory, oil field and shipyard has adopted its use almost exclusively for heavy rigging and haulage work, requiring transmission through space of proficient and dependable power.

The general care and proper handling of wire rope, while in the process of splicing, has always been considered the work of an experienced rigger. It will be found, however, that almost anyone with a little mechanical aptitude and energetic determination can execute a fairly good splice, after once mastering the proper methods that are used in each operation, as each step of the work is in itself quite simple to execute once the knack of tucking the strands and feel of handling the rope has been acquired. In fact the secrecy which surrounded the ability of a limited number of men who did this kind of work in former years was greatly exaggerated, as it has been proven among numerous small boat owners and yachtsmen of the present day that anyone

so inclined can do his own wire splicing with excellent results, for, as previously stated, a little experience in the proper use of the necessary tools and a small vise for convenience in handling the job properly will go far toward simplifying what otherwise appears to be a difficult job.

Every known type of wire splicing that the authors have come in contact with during many years of research, has been presented in this book. Included among the various examples are outstanding illustrations of the Liverpool, French and Logger's eye splice. Tail splices and both short and long splices are also shown in a number of clear drawings and photographs. Grommets are likewise described and illustrated. Lock tucks in various types of eye splices are explained here for the first time. The roll splice or Flemish Eye in wire will also be found of unique interest and will undoubtedly stir up great enthusiasm among yachtsmen as a new and simple method of forming an eye in wire rope. Multiple strand and single strand 1 x 19 stainless steel splicing in aerial and standing rigging wire have been amply covered with numerous illustrations.

Flexible and non-flexible cable splices that are used in aircraft cord are described in detail, as are also numerous other rare and unique examples, with an explanation of the uses for which each splice is intended.

Rope splicing has been used practically ever since rope was first made, as it was undoubtedly originated very shortly after the first form of rope was discovered, since rope alone, without some form of splice to go with it, is almost useless at times, while on the contrary if it is used in connection with the proper kind of splice it has almost unlimited possibilities in the performance of countless jobs.

The endless number of things for which various types of rope splices are used are so well known that they need but very little elaboration. However, suffice it to say that splicing of rope is in general use on all ships, and in shipyards for rigging and other purposes, and is also used extensively for ropes that are used in transmitting power in factories where leather belts or other means are not suitable.

Farmers and cowboys as well as seamen, and in fact men in almost every walk of life find occasions where a knowledge of how to splice rope is very handy and convenient. It is difficult to overestimate the importance of knowing how to do this work, since splicing is used in so many ways, that are essential for jobs in our daily life, when the ability to turn in a rapid and secure splice will often save an almost hopeless situation in an emergency.

It has therefore been the intention of the authors as far as possible to include every known variety and type of both wire and rope splice, many which never before have appeared in print on this specialized subject. A generous selection of blocks and tackles including handy-billys, and to our knowledge the first illustration on how to reeve two three-sheave blocks which are commonly found on lifeboat falls have also been added.

Every effort has been made to simplify the descriptive text as much as possible. The illustrations are presented in a manner that will clarify each step of the operation.

It need hardly be emphasized that the student or beginner should thoroughly master all details of each operation before attempting to master a difficult appearing job. Particular attention should also be paid to neatness and thoroughness in order to secure a safe and dependable splice.

Table of Contents

List of Illustrations

xiii

Splicing Wire and Fiber Rope

Wire Splicing

Wire rope, which has nowadays almost entirely replaced chains and fiber ropes for haulage and hoisting purposes, is made with a varying number of wires to the strand and a varying number of strands to the rope, according to the service for which the different ropes are intended and the degree of flexibility required.

There are five principal grades of wire rope manufactured, as regards the material from which the wires are drawn, viz.: Crucible steel, plow steel, extra strong crucible steel, iron, and a so-called blue-center steel. Occasionally copper and bronze wires are used in wire rope designed for light service and as a means of preventing corrosion. In so far as the flexibility of the rope is concerned there are but three commonly used classifications, viz.: Ordinary flexible, extra flexible, and special extra flexible.

Regular Lay of Wire Rope comprises the wires in the strands laid up from right to left with the various strands making up the rope laid up from left to right. This is also known as right lay rope; wires laid up to the left with the strands laid up to the right. Standard rope is made right lay.

In lang lay rope the wires in the strands and the strands in the rope are laid up in the same direction, either from left to right or the reverse. Lang lay rope is somewhat more flexible than standard rope, and as the wires are laid up more axially in the rope longer surfaces are exposed to wear, thereby increasing the endurance of the rope. Regular lay rope is the most commonly used since it hangs without twisting.

Classes of Wire Rope. Such wire rope as is used for haulage in mines, and around docks, usually consists of six strands of seven wires each laid up around a hemp core or center. Hoisting rope for elevators, derricks, mine lifts, and other similar purposes consists of six strands of 19 wires each wound around a hemp core. A more flexible rope for crane service and the like is made up of six strands of 37 wires each, wound around a hemp core.

In general the flexibility of the rope is increased by increasing the number of wires in each strand. Probably the most flexible rope made consists of six strands of 61 wires each. Other types comprise flattened strands for haulage, hoisting and transmission, non-spinning rope for the suspension of loads at the end of a single line, steel clad rope for severe conditions of service, guy and rigging rope and hawsers for towing and mooring.

Standard Types of Wire Rope are made up of 7, 12, 19, 24, and 37 wires each. Such a rope consisting of six strands of 12 wires each is commercially known as 6 by 12 rope, which is obtainable in varying degrees of flexibility, as previously explained, as are the other sizes.

When rope is used for ship's rigging, derrick guys, or under similar conditions involving continued exposure to the elements, the wires should be galvanized. Rope subjected to constant bending around drums and sheaves is not usually so treated.

Strength of Wire Rope. In determining the working strength of wire rope it was formerly the practice for each manufacturer to test the strength of each wire in the rope and then base the ultimate strength of the entire rope upon the number of wires and strands. However, the strengths arrived at in this manner were

3

usually in excess of the actual breaking strength of the rope. Today there are fixed standards for the strength of wire ropes of different sizes. In general a factor of safety of five is allowed in giving the working loads. These are given in the accompanying tables for four of the most commonly used sizes.

Although wire rope possesses great strength it must be used with considerable care if a full measure of service is to be obtained. In the manufacture of wire rope particular care is exercised to see that each wire in each strand and each strand in the rope is laid up with an equal amount of tension, so that when stresses are applied to the finished rope an equal amount of the strain will be carried by each component part of the rope.

Handling Wire Rope. Wire rope should not be coiled or uncoiled like fiber rope. If it is received in a coil it should be rolled upon the ground like a hoop and straightened out before being placed on a drum. If the coil is on a reel the latter should be placed on spindles or flat on a turntable and properly unwound. In any event every effort should be made to prevent the rope from kinking or untwisting.

Sizes of Drums. In like manner considerable care should be exercised in the choice of the size of the drums and sheaves. Whenever possible the diameter of the drum or sheave should be at least 700 times that of the smallest wire in the rope, and under no circumstances is it recommended to use a drum or sheave which is less than 300 times the diameter of the smallest wire. Care should be exercised too in the size and shape of the grooves in the drums and sheaves. These grooves should have a radius at the bottom slightly larger than that of the rope, their sides must be free from scratches or grooves and they should be so located upon the drum to permit of free running of the rope without its scraping on the sides or outer edges of the grooves.

Another consideration in the care of wire rope is the manner of reeving it around the drums and sheaves. In so far as is possible these should be so placed that there will be no reverse bending of the rope. This practice will wear out a rope quicker than any other abuse. A little care will usually eliminate circumstances which contribute to reverse bending.

Lubricating Wire Rope. Wire rope should be protected by a suitable lubricant, both internally and externally, to prevent rust and to keep it pliable. The lubricant used should not only cover the outer surface of the rope but it should also penetrate into the hemp center, to prevent it from absorbing moisture, and at the same time lubricate the inner surfaces of the wires and strands. Best results cannot be obtained from thick, heavy grease and oils and the sticky compounds frequently used for this purpose.

Seizings. Before cutting wire rope it is essential to place several sets of seizings around the rope on each side of the intended cut, to prevent disturbing the lay of the rope after the cut has been made.

It is important to use the proper grade and size of wire in making seizings such as annealed iron wire in the following sizes:

Size of Wire Rope (Diameter)	Size of Seizing Wire (Diameter)
⅜″ to ½″	.047″
⅝″	.054″
¾″	.063″
⅞″ to 1⅛″	.080″
1¼″ to 1⅞″	.105″
2″ and larger	.135″

Applying Seizings. The following instructions should be followed carefully in applying seizings to wire rope. The wire should be wound uniformly and firmly. Unless a serving mallet is used there is no advantage in making more than ten wraps of the wire for each seizing.

After the seizing has been applied and the wire pulled up taut, cross the ends of the wire over the seizing and twist the wires counterclockwise. Grasp the ends of the seizing wire with wire cutters and twist up the slack. Do not attempt to tighten up on the seizing by twisting its ends. Cut off the ends and hammer the twisted end of the seizing back against the standing part of the rope.

Number of Seizings Required. Two seizings are necessary on iron rope, three on steel rope, and four should be used on independent wire rope center ropes, while for larger ropes even more seizings should be used.

Splicing Wire Rope. The splicing of wire rope is usually entrusted to workmen who possess some degree of mechanical skill and ability in handling tools. It follows that the greater the degree of the skill of the workman and the care employed the more satisfactory will be the result.

It would, therefore, be well for those who are entirely lacking in experience to make several practice splices before attempting to splice a wire rope subject to severe conditions in actual use.

Plates 1 and 2—A Liverpool or Spiral Eye Splice

FIG. 1A: *The Liverpool or Spiral Splice, First Method.* To make this splice involves the use of worming, parceling, and serving, and the application of a metal thimble in the eye of the splice. The first step is to determine the amount of rope required to completely encircle the thimble. Mark this length off on the rope allowing enough rope, about 18 inches, in addition, to form the splice. That part of the rope which will be enclosed by the thimble is next coated with a good preservative, after which it is wormed, parceled and served as shown on PLATES 6 and 7, FIGS. 9B and C.

Next bend the rope to form a bight in which to place the thimble and clamp both the rope and the thimble in a vise in the manner shown in the illustration. Proceed to unlay the strands of the short end and seize the end of each strand. Cut out the core of the rope close to the serving around the eye.

To prepare the standing part of the rope for splicing insert a marline spike under two strands of the rope, making certain that the spike is inserted with the lay of the rope and not against it. The spike is inserted under the strands close to the thimble and then rolled out to the position shown. This is done to open up the strands because a strand of wire cannot be tucked with a short bend in the strand as is done with fiber rope, which is the reason for the longer opening between the strands.

With the marline spike in place the strand *a* is then inserted under the two strands, working the strand *a* from the point of the spike toward the handle. With strand *a* in place it is pulled up taut and then rolled back toward the thimble until it occupies the position shown in FIG. 1B. The method employed in rolling the different strands back into place is shown in the views at FIGS. 1B and C.

B: As the next step one strand of the standing part of the rope is raised with the marline spike in the manner previously explained. Strand *b*, the strand next to *a*, is placed in this opening and is in a like manner rolled back toward the thimble. This procedure is followed with the remaining strands in succession, tucking each strand of the end under one strand of the standing part.

C and D: Other operations in completing the Liverpool Splice are shown in the

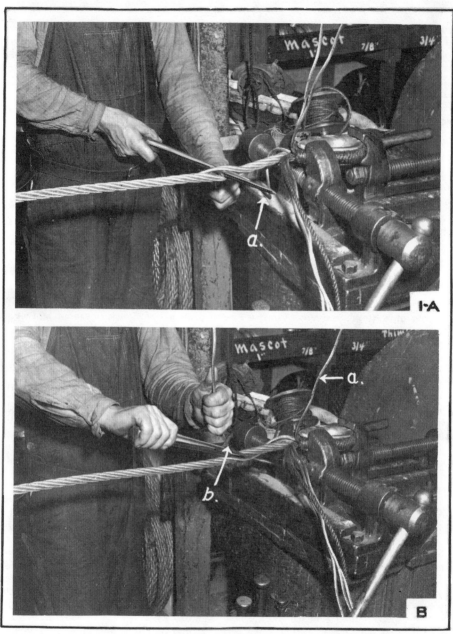

PLATE 1—A LIVERPOOL OR SPIRAL EYE SPLICE

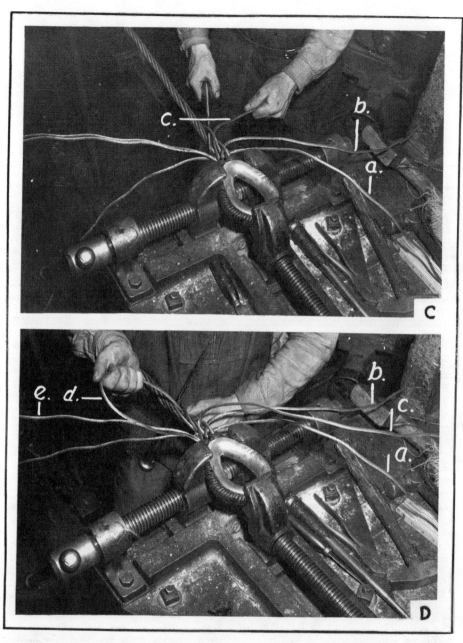

PLATE 2—A LIVERPOOL OR SPIRAL EYE SPLICE *(Continued)*

accompanying illustrations, in their various stages as the work progresses.

The extra length of rope, required for leaving sufficient margin to go around the thimble and for making the splice, varies as shown in the following table:

Diameter of rope in inches	1/4–3/8	1/2	5/8–3/4	7/8–1	1 1/8	1 1/4	1 1/2	1 3/4	1 7/8	2
Extra length to allow in feet	1	1 1/2	2	2 1/2	3	3 1/2	4	4 1/2	5	5 1/2

Plates 3 and 4—A Rigger's Bench and Tools

FIG. 1E: *The Liverpool Splice,* First Method continued. In this illustration strand *e* has been tucked into place and the strand *b* is shown rolled into its final position. The work is continued by tucking all of the strands again, in the same order, taking care to tuck them over and around the strands they were first tucked under. In this manner three more rounds of tucks are made with all of the strands, making four rounds of tucks in all. Strands *a, b,* and *c* are given one more tuck each, making four and one-half tucks, which is sufficient for ordinary use.

If it is desired to taper the splice this is done in the following manner: After having made four and one-half tucks separate each strand into halves. Tuck one half of each of the half-strands twice more and then halve the strands again and proceed to again tuck one-half of each of the half-strands once more.

The splice is then beaten well, beginning at the thimble, to lay the various strands more correctly in position. The manner in which the strands are tucked in making eye splices is shown more clearly on PLATE 8, FIGS. 10A to G.

FIG. 2: *A Wire Net Cargo Sling.* Nets or slings such as these are employed in such work as is too severe for the ordinary rope sling. The frame ropes *b* may be in one piece with the ends short-spliced together, or they may be made up of four separate pieces of rope with eye splices made in all four corners of the net. The frame ropes are usually larger in diameter than those used for forming the mesh of the net.

The mesh ropes *a* are put in by unlaying two strands from a length of rope. Then begin by laying up these two strands again and each time one of these ropes intersects one of the other mesh ropes running at right angles to it the two unlayed strands are tucked under two of the strands of the cross-rope. This method is continued until the frame on the opposite side of the net is reached. Both ends of the cross-rope are then spliced into the frame-rope. The Beckets *c* may be made of either wire or fiber rope, and as may be seen from the illustration they are joined to the corners of the frame-rope with eye splices in one end while a metal ring for the hoist is spliced in the other end.

FIG. 3: *A Rigger's Vise.* This is one of the tools that is very essential in splicing wire rope. It acts both as a press for forming the splices in the ends of a rope and at the same time it acts as a vise for securing and holding the rope while the splice is being made. After the bight for the eye has been formed in the end of the rope, it and the thimble are placed between the jaws of the vise. The jaws, operated by screw threads on a bar passing through them, clamp the rope and thimble and hold

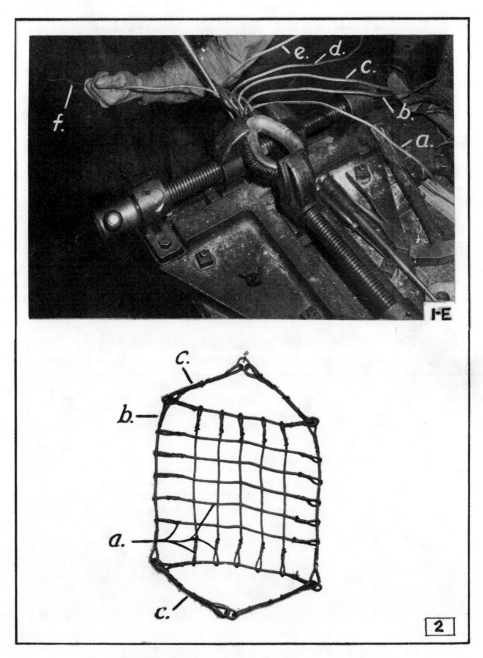

PLATE 3—A LIVERPOOL SPLICE AND CARGO SLING

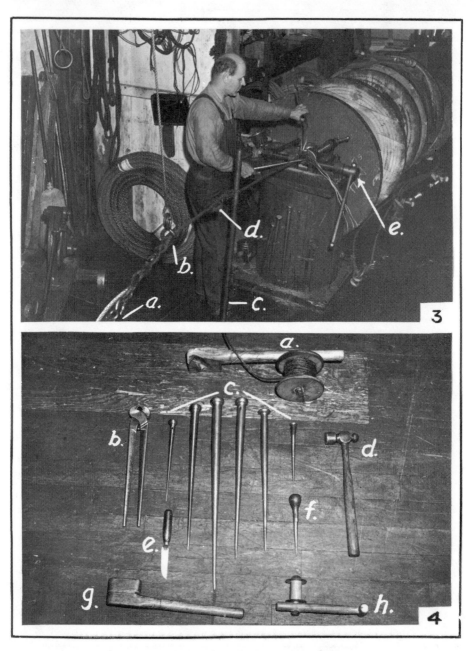

PLATE 4—A RIGGER'S BENCH AND TOOLS

them firmly. This threaded bar is shown at *e*. In putting the rope into the vise the end should be on the right with the standing part on the left next to the workman. Sufficient material should always protrude from the vise jaws to allow for forming the splice.

After the rope and thimble have been clamped in the vise a strop is placed around the rope and a block and tackle is attached to this, as at *a*. A pair of sister hooks, *b*, which are suspended from the ceiling, are then placed on the rope, between the vise and the block and tackle, *a*. The next step is to take up on the block and tackle until the wire rope is pulled taut. When this is done another strop *d* is placed on the rope, between the sister hooks and the vise. A bar *c* is then placed through the bight of the strop *d* and several turns are taken with the bar about the rope, to take out some of the turns in the lay of the rope. It will be found that the use of this bar aids materially in untwisting the lay of the rope in order that the marline spike may be more easily inserted between its strands.

Also, when a splice is made in a wire rope held under tension as explained, the work is easier to perform and at the same time the splice presents a neater appearance.

FIG. 4: *Tools Used in Splicing Wire Rope.* Three serving mallets are shown at *a, g,* and *h*. PLATE 7, FIG. 9C, shows how serving mallets are used on wire rope. The tool shown at *b* is a pair of nippers or wire cutters such as are used for cutting the wires of the various strands of the rope and the wires of any seizings that may have been used. The slender pointed implements shown at *c* are various forms of marline spikes. These are used for separating the strands in the process of making splices. A ball-peen hammer is shown at *d*. These are typical mechanic's hammers common to the mechanical trades. A knife such as that shown at *e* is used for the various operations of cutting the core of the rope or for trimming Marline and serving materials. The pricker shown at *f* is used as a marline spike in splicing the smaller sizes of ropes.

Plate 5—A French Eye, and a Logger's Eye Splice

FIG. 5: *The Liverpool Splice,* Completed. The Liverpool Splice illustrated on PLATES 1 and 2, FIGS. 1A to D, and on PLATE 3, FIG. 1E, is shown here as it appears after having been wormed, parceled and served.

FIGS. 6A to C: *The French Lock Eye Splice with a Thimble.* The method employed in making this splice is essentially the same as that followed in making the Liverpool Splice. The first set of tucks is made in precisely the same manner as illustrated on PLATE 8, FIGS. 10A to 10F, after which the strands *a, b,* and *c* are given one additional spiral tuck each, as shown on PLATE 8, FIG. 10G.

The next step, shown on PLATE 5, FIG. 6B, is to tuck each strand over one and under two, against the lay of the rope until three more tucks have been made. There will then be four and one-half tucks. The core of the rope, which was pushed down beside the eye, is then cut off short and the strands are each cut off, after the splice has been beaten into shape with a hammer or mallet. The splice will then appear as shown in FIG. 6C.

This type of splice is used in ropes on which the load hangs free and might have a tendency to untwist the lay of the rope, thereby allowing the strands forming the splice to slip or pull out, such as might be the case with a Spiral or Liverpool Splice.

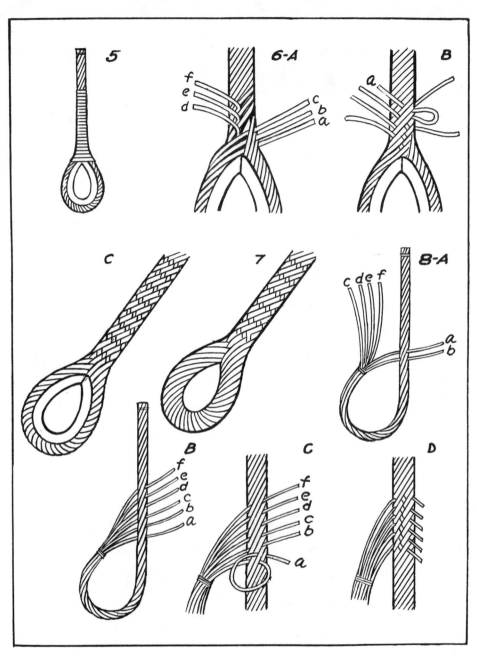

PLATE 5—A FRENCH EYE, AND A LOGGER'S EYE SPLICE

Fig. 7: *The French Lock Eye Splice Without a Thimble.* It is made in precisely the same manner as that previously explained. It should be understood in this connection that an eye splice with a thimble is considerably stronger than one without and that all of the wear is placed directly upon the rope instead of being absorbed by the thimble.

Figs. 8A to D: *The West Coast or Logger's Splice.* This splice is practically the only type used by loggers and stevedores on the Pacific coast. Although it may appear to be weaker than other types the authors have never heard of a case in which the eye in such a splice pulled out or slipped. It is one of the simplest forms of wire rope splicing.

As the first step cut out the core close to the seizing on the end. The strand *a* is then tucked under two strands of the rope.

Next, the strand *b* is passed under the following two strands in the rope. This process is repeated until all six strands of the splice have been passed successively under two strands each of the rope.

After the first round of tucks has been made, that is, after all six strands have been tucked, the work will appear as shown in Fig. 8B. The next step, taking the strands in their respective order from *a* to *f*, is to pass each strand back to the left over one strand and under two of the rope, with each strand tucking over one to the left and under two to the right, as shown in Fig. 8C in the accompanying illustration.

This completes the splice, except that it is beaten with a hammer or mallet and the strands are cut off. They should not be cut off too short, however, as it is better to allow them to extend through the rope about one-half to three-quarters of an inch. The finished splice is shown in Fig. 8D.

Plates 6 and 7—Worming, Parceling, and Serving

Figs. 9A to D: *Worming, Parceling, and Serving a Splice.* Some riggers when starting to make a splice take the first tucks in the manner shown in Fig. 9A, that is, one strand under two strands, against the lay of the rope. This method does make a neater job. The other tucks are then made in the same manner as was explained in making the Liverpool Splice.

The next step, after all the tucks have been made and the splice has been beaten out, is to worm the grooves between the strands of the rope with a strand of spun yarn. This is represented by *a* in Fig. 9B. The splice is then parceled with a strip of burlap or tarred canvas, shown in Fig. 9B at *b*.

After the worming and parceling have been completed the entire splice is then served over the parceling with spun yarn. This is applied with a serving mallet as shown at *c* in Fig. 9C. Notice that the parceling is applied by starting it at the thimble and working it out along the rope to the end of the splice, while the serving is applied by starting it at the end of the splice and working it toward the thimble. This is more aptly explained in the following adage:

> Worm and parcel with the lay,
> Turn and serve the other way.

The completed splice is shown in Fig. 9D. In making a splice, the lay of the strands is disturbed; to protect the rope from moisture, the splice should be wormed, parceled and served carefully.

The rope is wormed in order to fill out the grooves between the strands of the rope. The parceling is applied in order to keep the moisture out of the Splice and also to make a smoother and even surface upon which the parceling is applied. The parceling is put on of course to secure the whole job, which is shown in Fig. 9D.

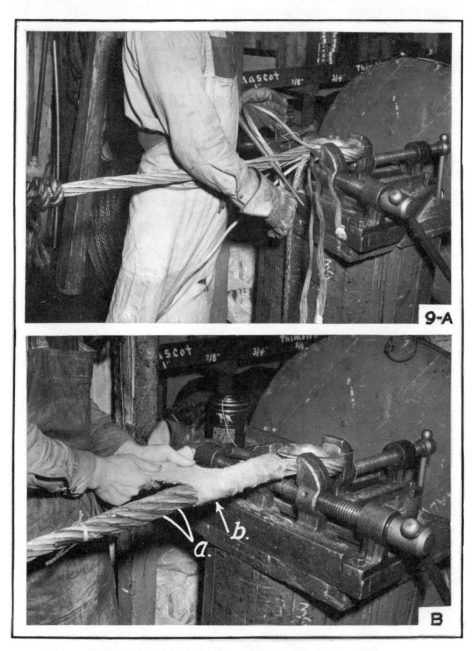

PLATE 6—WORMING, PARCELING, AND SERVING

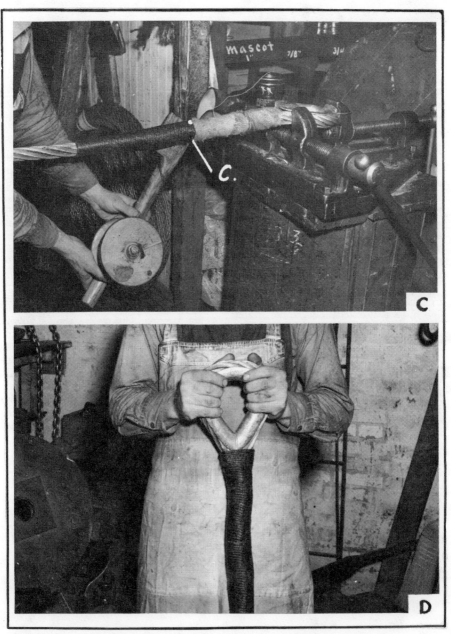

PLATE 7—WORMING, PARCELING, AND SERVING *(Continued)*

Plate 8—Diagrammatic Steps in Splicing and Seizing

FIGS. 10A TO G: *The Liverpool Splice,*
Second Method. The Liverpool or Spiral
Splice is the method most commonly used
in splicing wire rope. Not only can it be
made rapidly but at the same time it makes
a neat, strong and dependable splice.

As explained in the First Method, PLATE
1, FIG. 1, a rigger's vise was used for holding the work. When no such vise is available an ordinary mechanic's vise may be
used, but in doing so the thimble must
be held in the eye by seizings as shown at
b in FIG. 10A. The rope itself should also be
seized about one and one-half feet from its
end. This end seizing is placed at *a*, FIG.
10A, when seizing the thimble in the bight
forming the eye.

All of the strands of the end of the rope
are then unlaid back to this seizing, a marline spike is inserted under three strands
of the rope and the strand 1 is tucked as
shown in FIG. 10A. As the next step the
spike is withdrawn from the rope and again
inserted in the rope but this time under
two strands, after which strand 2 is tucked
as shown in FIG. 10B. After this operation
has been completed the spike is again inserted in the rope, but under only one
strand, and the strand 3 is tucked. This is
shown in FIG. 10C.

As may be noticed in this view, FIG. 10D,
the splice has been turned completely over
from left to right, hence, strands 1, 2, and
3 now appear in back of the standing part
of the rope. This was done in order to
more clearly explain the following steps,
but is not necessary in making a splice of
this kind.

Continuing the work the spike is inserted under the next strand to the left of
the strand under which strand 3 was
tucked and strand 4 is tucked in alongside

the spike. This is followed by lifting the
next strand of the rope and tucking strand
5 in alongside the spike. The same procedure is followed in tucking strand 6. These
operations are clearly shown in FIGS. 10E
and 10F.

After all of the strands have been tucked
in the manner explained the work is continued by again starting with strand 1 and
tucking it under the next strand of the
rope. Notice, however, that in the following successive steps the strands are tucked
around and around the strands in the
standing part which they were previously
tucked under, that is, the tucks are made
from left to right, as shown in FIG. 10G.
This is commonly known as the Spiral
Tuck.

FIGS. 11A and B: *Applying Wire Seizing,*
First Method. Seizings are applied to wire
rope in the same manner and for the same
purpose they were used on fiber rope; to
prevent the strands of the rope from becoming unlaid. It should be remembered
also that before a wire rope is cut seizings
should be clapped on the rope on both
sides of the intended cut. The correct number to use for ropes of different kinds was
given in the introductory paragraphs to
this chapter.

The manner in which a serving mallet is
used and the finished seizing are shown
in FIGS. 11A and 11B, respectively.

FIGS. 12A, B, and C: *Applying Wire Seizing,* Second Method. These three illustrations show the progressive steps in applying wire seizing when no serving mallet is
used. The wire is wound first as shown in
FIG. 12A: the ends are twisted as in FIG.
12B, after which they are cut off. The finished seizing is shown in FIG. 12C.

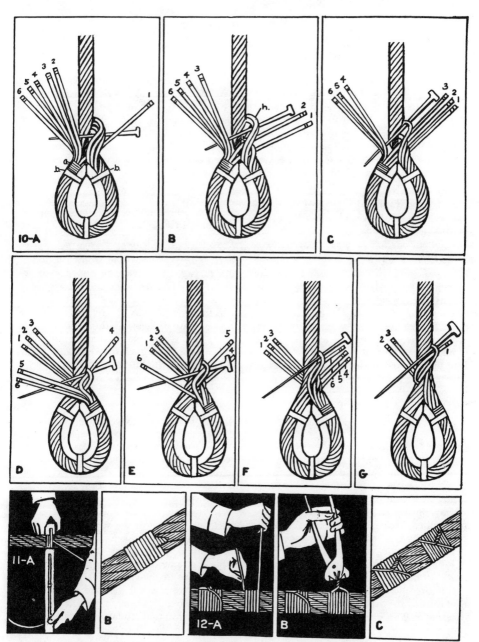

PLATE 8—DIAGRAMMATIC STEPS IN SPLICING AND SEIZING

Plate 9—The Lock Tuck in a Liverpool Eye Splice

FIGS. 13A and B: *The Liverpool or Spiral Splice* as it appears after having completed the first steps previously explained (*see* PLATE 8, FIG. 10). When the work has advanced to the stage shown in the accompanying illustration three more rounds of tucks are made. The splice may or may not be tapered as desired. That shown here is not. Finished splice is shown in FIG. 13B.

FIG. 14: *A Finished Liverpool or Spiral Splice* in which four and one-half rounds of tucks were made.

FIGS. 15A, B, and C: *The Liverpool Eye Splice with a Lock Tuck.* The strand *a* is tucked under two strands against the lay or from right to left. Next proceed as in other splices to tuck the first strand *b* under two and each following strand under one as in PLATES 1 and 2. When strands *c, d, e,* and *f* have been tucked turn the splice around. Strand *a* will then appear in front. It is then tucked under the next strand in the standing part to the left of strand *f*.

The view at FIG. 15B illustrates the manner in which strand *a* is tucked. The other strands are tucked in the same manner as was used in making the Liverpool Splice shown on PLATE 8, FIG. 10.

Three and one-half additional tucks are taken and the strands are cut off to finish the splice. This form of splicing is employed when it is desired to make a closer and neater job, especially in that part adjacent to the thimble. The finished splice is shown in FIG. 15C.

FIGS. 16A and B: *The French Eye Splice with a Lock Tuck.* The strands *a* and *b* are tucked under two and one strands respectively. The strands *c* and *d* have their positions reversed as may be seen in the illustration. Strands *e* and *f* are to be tucked with the lay of the rope. Strands *a, b,* and *c* are each tucked again, after which proceed to tuck over one and under two as shown in PLATE 5, FIG. 6B.

The finished splice is shown in FIG. 16B.

Plate 10—A Long Splice, and a Tiller Rope Splice

FIGS. 17A, B, and C: *The Long or Endless Splice.* It is extremely important, in the making of a long splice in wire rope, to use great care in laying the various rope strands firmly into position. If, during any of the various operations, some of the strands are not pulled tightly into their respective places in the finished splice, it is doubtful if satisfactory results will be obtained.

When such a splice is placed in service those strands which are relatively slack will not receive their full share of the load, thus causing the other strands to be stressed excessively. This unbalanced condition will result in a distorted relative position of some of the rope strands so that they will be projected above the normal diameter

of the rope and consequently will be subjected to abnormal abrasion and abuse. In addition, the unequal stress distribution will decrease the possible ultimate strength of the splice.

It is strongly recommended, therefore, that during each of the steps explained in the following method, particular attention be paid to maintaining as nearly as possible the same degree of tightness in all of the strands in the splice.

When ropes are to be used in places in which their failure may result in material damage or might endanger human lives the splicing should be done only by men who are well experienced in this work. It is considered good practice with such splices

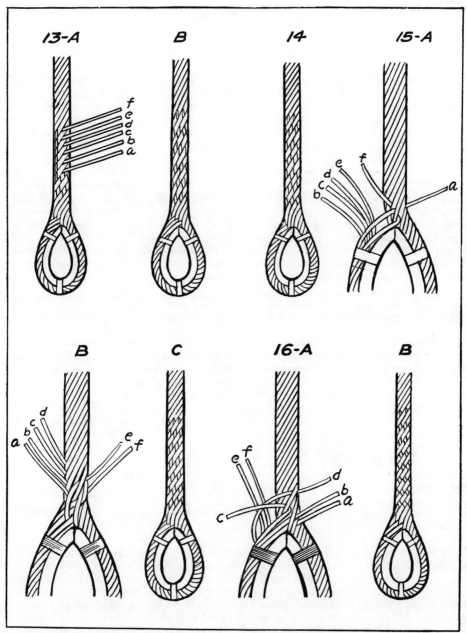

PLATE 9—THE LOCK TUCK IN A LIVERPOOL EYE SPLICE

as these to test them under stresses equal to at least twice their maximum working load before the ropes are placed in service.

As an aid in making long splices the accompanying table shows the amount or length of rope to be unlaid on each of the two ends of the ropes for ropes of different diameters. It will be noticed that the data given in the table are for regular lay wire rope. If lang lay rope is to be spliced doubling the lengths of the splices is recommended.

The instructions included in the following explanation are for making a 30-foot splice in ¾-inch diameter rope. The first step is to determine the total or overall length desired in the rope after the splice has been made, bearing in mind in measuring this length that an additional length of 15 feet will be required in each piece of rope, to be used in making the splice.

After the rope has been cut and a seizing applied 15 feet from the end of each rope, unlay the strands of each section up to the seizing. Cut off the cores close to the seizing and then clutch or marry both ends in such a manner that the various strands of each section will interlace with the corresponding strands of the other section.

After both ropes have been married they are pushed together as tightly as possible, until the cores butt against each other. Another seizing is placed on the ropes at the point where they join; this will serve to hold both ropes firmly together. The two original seizings which were put directly on the ropes are now removed. Lift one of the unlaid strands of the rope on the left and put a stout seizing on the remaining five strands. Remove the last seizing put on both ropes at the point where they joined. Next unlay the strand just lifted to the left for a distance of thirteen feet. Now fill this groove with a strand from the rope on the right. When this has been done there will remain only two feet

of this follow up strand as shown at *a* in Fig. 17A. The long strand *b* is then cut, leaving only two feet of the end remaining.

The strands remaining in the center are unlaid from one rope and are followed up with the strands from the other rope. Care should be exercised in laying up the strands from the two ropes in that alternate strands are brought into their proper positions. That is, one strand from the rope to the right is used to replace its corresponding strand in the rope from the left. In other words, three strands are laid up from the rope to the right with the three corresponding strands from the rope to the left, each spaced equally distant apart.

The ends of the strands must now be secured without increasing the diameter of the rope. The first step is to straighten out all of the ends and then wrap each strand with a length of friction tape, making the strand of the same diameter as that of the core of the rope.

When this has been done take a sharp knife and cut the fiber center of the rope at the point where both strands meet, and pull it out to the right for a short distance. Insert a marline spike under one strand and at the same time strand *b* is worked into the center of the rope in place of the heart *h* that has been pulled out. Continue to work the strand *b* into the rope as the core is pulled out until all of the wire strand has been worked into the rope in place of the core. Cut the core off at the end of strand *b* and work the remaining part of the core back into the rope center.

This same procedure is again carried out in order to dispose of strand *a* but working to the left this time rather than to the right as for strand *b*. There now remain five more sets of strands to be tucked. These are disposed of in exactly the same manner as has just been outlined. The rope is hammered well at the points where the strands enter the rope in order to give a smooth and neat appearance to the finished splice.

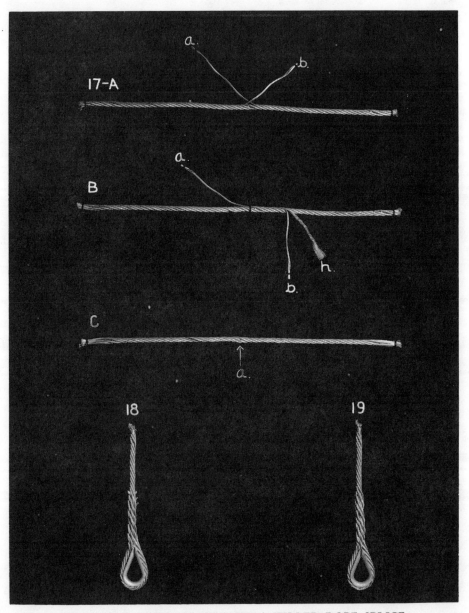

PLATE 10—A LONG SPLICE, AND A TILLER ROPE SPLICE

The finished splice is shown in FIG. 17c, the symbol *a* being the point at which the two strands cross. (*Also see* FIGS. 32A to F, PLATE 15.)

The total amount of rope to allow for making both short and long splices is shown in the following table:

Rope Diameter (Inches)	Length to Allow (Feet)	
	Short Splice	*Long Splice*
¼– ⅜	15	30
½– ⅝	20	40
¾– ⅞	24	50
1 –1⅛	28	60
1¼–1⅜	32	70
1½	36	80

The total length of tuck to allow in making both short and long splices is shown in the following table.

For six-strand rope, the length of the tuck is approximately one-twelfth the amount of rope allowed for the splice; for eight-strand rope, it is approximately one-sixteenth the amount allowed for the splice. Both dimensions are in inches.

Rope Diameter (Inches)	Length of Tuck (Inches)	
	Short Splice	*Long Splice*
¼– ⅜	15	30
½– ⅝	20	40
¾– ⅞	24	50
1 –1⅛	28	60
1¼–1⅜	32	70
1½	36	80

FIG. 18: *The Liverpool Splice in Tiller Rope.* This kind of rope is usually made of copper or bronze wire and at times of galvanized wire laid up left handed. The splice is made exactly as with right hand rope except that the various steps are reversed.

FIG. 19: *The French Eye or Lock Splice in Tiller Rope.* It is made in the same manner as the splice shown on PLATE 5, FIG. 6, except that it is worked to the right rather than toward the left.

Plates 11 and 12—A Wire Rope Socket and Its Application

FIGS. 20A to D: *Attaching a Socket to Wire Rope.* Sockets of the proper construction, if carefully and properly attached to a wire rope with melted spelter form a very effective method for attaching the ends of such ropes to other objects.

The first step in attaching a socket is to seize the end of the rope as at *a*. The material used for this purpose is soft iron wire such as that used for ordinary seizings. After the end has been seized the strands are opened up and the core is cut off short. Next, separate the wires of each strand and thoroughly clean each of them in preparation for applying the spelter.

To clean the wires they should be washed thoroughly in either gasoline or kerosene and wiped dry. Next, dip the frayed wires only into a solution of one-half commercial muriatic acid and one-half water, from thirty seconds to one minute, or until the acid has thoroughly cleaned each of the wires. (Use extreme precautions to prevent the acid from coming into contact with the rope strands.) Remove the wires from the acid bath and dip them into a solution containing soda, to neutralize the effect of the acid, after which dry the wires thoroughly. In making the acid bath the solution should under no circumstances be stronger than one-half acid and one-half water, otherwise serious damage might be done to the wires.

Continue the work by compressing the

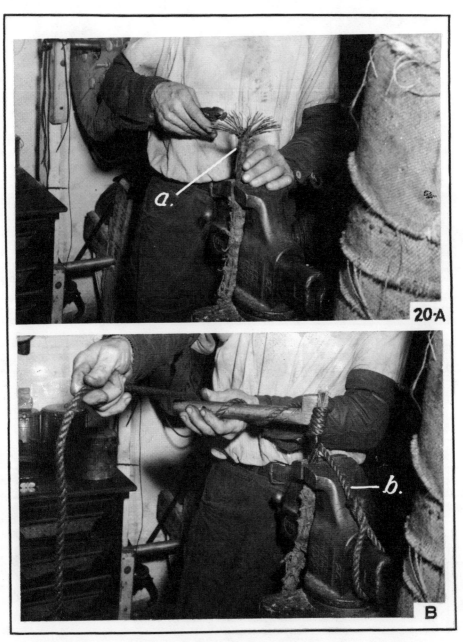

PLATE 11—A WIRE ROPE SOCKET AND ITS APPLICATION

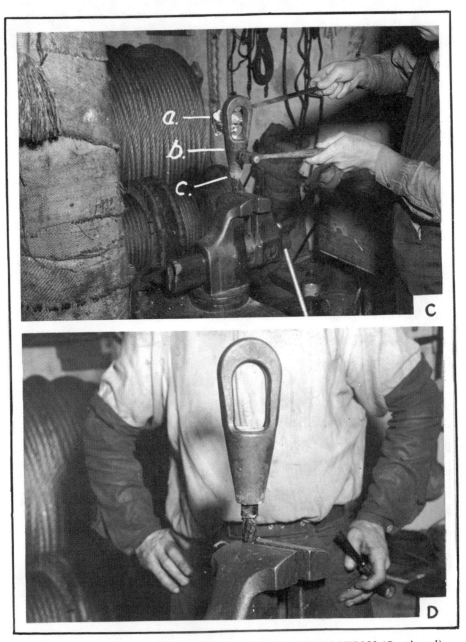

PLATE 12—A WIRE ROPE SOCKET AND ITS APPLICATION *(Continued)*

wires together until they are of such diameter that the socket can be slipped down over them. This may be done by the aid of a length of fiber rope and a serving mallet, as shown in Fig. 20b. The rope has an eye in one end which is passed over the horn of the vise, after which the free end is served about the frayed wires with sufficient tension applied to them to force them back into a small cylindrical mass. As the socket is slipped down over the compressed wires the rope serving is slowly unwound until the socket can be forced down to the previously placed seizing.

As the next step in the work a small wad of fire-clay or asbestos cement is placed around the bottom of the socket to retain the molten spelter, as shown at c in Fig. 20c. The entire end of the rope as well as the socket is then heated with an oil or gas flame to a temperature equal to or

slightly above that of the molten spelter. Care should be exercised in the heating to prevent the heat from becoming great enough to affect the temper of the wires in the rope.

The individual wires of the rope should not be allowed to extend above the top of the basket or pocket-like opening in the socket and in pouring the spelter a ladle of sufficient capacity should be used to permit of completely filling the socket in one operation. Care should be exercised too in aligning the socket with the rope and as the spelter is poured the socket should be tapped lightly with a hammer to insure that the spelter fills all of the small spaces in and around the wires.

The finished job is shown in Fig. 20d after any excess spelter and the fire clay around the bottom of the socket have been removed.

Plate 13—Splicing Multiple Strand Wire Rope

Figs. 21a and b: *Splices in Ropes of More than Six Strands*. In some types of rope used principally for special services the ropes are composed of more than six strands and these are laid up differently than is the practice with regular right or left laid rope. In one of the types of special wire rope there are 18 strands. These are divided into inner and outer groups which are both laid up one over the other around a fiber rope core. The arrangement of the strands is that six of them are laid directly around the core while the other 12 strands form an outer layer around the six strands and the core. In other words the inner six strands are laid up as in regular rope while the remaining 12 strands form an outer layer around it.

In making an eye splice in rope of this kind it is seized and placed around a thimble in the usual manner. After this has been done the strands are divided as follows: The 12 strands in the outer layer are separated into six groups of two strands each, in the standing part of the rope only, the core and its six surrounding strands being left intact. In the working or splicing end of the rope all of the 18 strands are used. These are divided into groups of three strands each. Then, as the work of splicing progresses, three strands in the splicing or working end of the rope are tucked under two strands in the standing part of the rope. This is clearly shown in Fig. 21a.

Before starting to make the splice the core of the rope is cut off short near the thimble and all of the working strands are laid out, care being taken that none of them cross over each other. The work of tucking the strands is done exactly as in making a Liverpool Splice, until all six of the outer strands of three strands each have been tucked under the six groups of two strands each in the standing part of the rope.

The next step is to tuck each group of three strands each in the outer layer over one group of two strands in the inner layer of the standing part and then under one group of two strands, working toward the left or against the lay of the rope. This completes the second round of tucks. The third and fourth rounds are made in the same manner as just explained.

Before starting to make the fifth round of tucks turn back one strand each in all of the six groups of three strands each in the outer layer, leaving two working strands in each of the six groups. Proceed then to tuck these groups of two strands each over one group of two strands in the standing part and under one group of two. This is repeated to complete the sixth and last round of tucks.

After all of the tucks have been made beat the splice well with a mallet and cut off the ends of the six strands, taking care to secure the ends well. The completed splice is shown in Fig. 21b.

Fig. 22a: *Splicing Single Strand Cable.* This kind of wire rope is not generally used in commercial work, although some of it is used in airplane construction. It is usually constructed with the outer wires laid up left handed and varies from 7 to 37 wires per rope. It is twisted 18 strands around 12 strands, around 6 strands, around one strand, the latter forming the core. In a rope of 19 wires these would be laid up or twisted 12 strands of wire around 6 and the 6 around one.

In the accompanying illustration the rope is shown with the thimble turned in ready for splicing. In this case the single wire core is cut off at the thimble and is not used in making the splice, thereby leaving 18 wires to be worked into the splice.

These are divided into groups and the work of splicing is carried out exactly as explained in Fig. 21a. Remember, however, that this is left laid rope and not right lay, making it necessary to make all of the tucks in the reverse manner to that used for right laid rope.

In Fig. 22b the first round of tucks is shown completed and the finished splice is shown at Fig. 22c.

Fig. 23: *Splicing Single Strand Rope of Thirty-Seven Wires.* Such rope as this is laid up 18 wires around 12 wires, around 6 wires, around one wire. The six inner wires and the single core wire are not used in making this splice but are cut off short at the thimble, leaving 30 working wires, which are divided into six groups of five wires each.

The splice is made as has been explained, except that the wires in the standing part of the rope are divided into groups of three wires each and that each of the five-wire groups is inserted or tucked under nine wires in the standing part, or three groups of three wires each. After the first row of tucks has been made the groups of five wires are tucked under two groups of three wires and in the next pass under one group of three wires.

After four full rounds of tucks have been made in this manner bend back two of the wires in each five-wire group and tuck three wires under three wires. In the next round bend back one wire and tuck two wires under three, taking two more turns with the remaining two wires in each group, going over three and under three in the standing part. The splice is then beaten out in the usual manner, after which it is good practice to put a small wire serving at the end of the splice to cover the ends of the wires.

Figs. 24a and b: *A Wire Rope Grommet.* These are made from a single strand of wire rope, which is laid up around a fiber rope core. The ends are finished off in

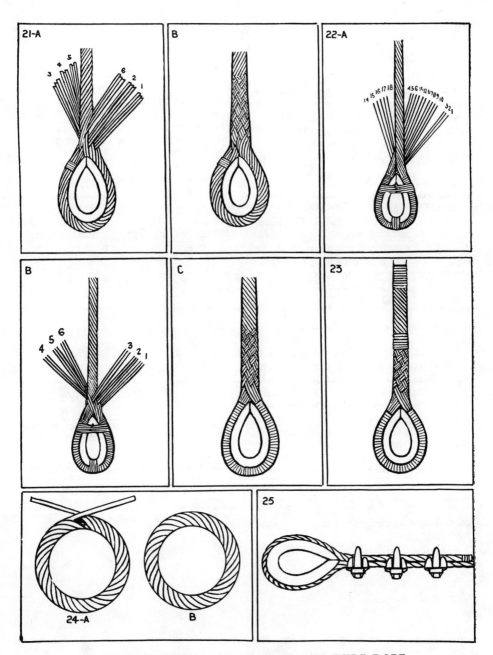

PLATE 13—SPLICING MULTIPLE STRAND WIRE ROPE

the same manner as was used in making the Long Splice, shown on PLATE 10, FIG. 17.

FIG. 25: *Wire Rope Clips.* These are used as shown to form an eye in the end of a wire rope. This is not the most satisfactory method of forming an eye and is frequently used in an emergency or when it is not possible to make one of the standard types of eyes.

The number of clips to be used and the distance they should be spaced apart is determined by the size of the rope and the service for which it is intended. The number of clips may vary from 2 to 8 for different sizes of ropes, while the length of the short end of the rope extending back from the thimble may be as much as from 35 to 50 times the diameter of the rope.

The illustration shows 5/8-inch diameter rope with three clips spaced 5½ inches apart. In the case of 1-inch rope the length of the end would be from 35 to 50 inches, which would require from six to eight clips.

To apply the clips place the thimble in the bight of the eye and seize the rope temporarily at the end of the thimble. Clap on the first clip farthest from the thimble and pull it up tight. Be certain that the base or saddle of the clip rests upon the standing part of the rope and that the U-shaped member is applied over the short end of the rope. This applies to all clips. The next step is to apply the clip nearest to the thimble, but it is not pulled up tight. If more than two clips are to be applied each of them should be clapped on equally distant from each other.

In applying clips in this manner it is advisable, whenever possible, to apply tension to both members in order to equalize the stresses in both, after which all of the clips may be pulled up tight. After the clips have been in service for short intervals the clips should again be tightened as the strains to which the rope may be subjected tend to work them loose.

The number of clips used with various sizes of wire rope is shown in the table which follows:

Diameter of Rope	Number of Clips	Space in Inches Between Clips
5/8	4	3¾
3/4	5	4½
7/8	5	5¼
1	5	6
1⅛	5	7
1¼	6	8
1⅜	7	9
1½	8	10
1⅝	8	10
1¾	8	11
1⅞	8	12
2	8	12

The following table gives the efficiency of thimble spliced eye attachments:

Diameter of Rope	Efficiency
¼" and smaller	100%
⅜" to ¾"	95%
⅞" to 1"	90%
1⅛" to 1½"	80%
1⅝" to 2"	75%
2¼" and larger	70%

Plate 14—A Tail Splice and a Short Splice in Wire Rope

Fig. 26: *The Tail Splice.* Such a splice as this is used for splicing fiber rope tails to the ends of wire rope. Usually they are made on the flexible wires of the running rigging such as on sailing yachts and small sailboats.

In the accompanying illustration, for the purpose of illustrating the work, a piece of ¼-inch rope is shown spliced to a length of ½-inch diameter manila rope. The manila rope is shown at *a* and the wire rope at *b*.

To start the splice the wire rope is unlaid back about 12 inches, after which three of its alternate strands are unlaid for another 12 inches, where a seizing is clapped on. The next step is to unlay the manila rope back 24 inches, after which the first three strands of the wire rope are married to the three corresponding strands of the manila rope. Lay up the strands of the two ropes again, covering the three strands of the wire rope. There will now be 12 inches of manila rope with three strands of the wire rope as a heart or core. There will also be three strands of the wire rope each protruding at two different points. The strands are now unlaid and with the aid of a sailmaker's palm and a sail needle each individual wire is stitched through the strands. Some workmen prefer to dispose of the strands intact without unlaying them. When this is done a special tubular needle is required. They are tucked around similar to a short splice in manila rope, but instead of going over and under they are stitched right through the manila strands. In either method the wires are completely covered.

In the illustration the strand *d* is being tucked, strand *e* has been tucked, while strand *c* is to be unlaid and stitched into the rope. The strands *f* still remain to be tucked into the rope. The ends of the manila rope at *g* are scrape-tapered, after which the entire splice is marled and served with sail twine.

Fig. 27: *A Bulb on a Wire Rope End.* This method is used in securing a socket to the end of a wire rope instead of attaching it with spelter. A seizing is clapped on the rope a short distance from its end, the strands are unlaid and straightened out, after which each strand is bent back individually over the seizing. If the rope has a wire core it too must be bent back with the other strands, but if the core is made of fiber rope it is cut off short at the seizing.

After the wires have all been bent back they are hammered down to form a solid wire bulb. The next step is to cut off each wire so as to form a taper similar to that in the socket. The taper is formed by cutting the ends of the wires at different intervals from their ends back to the end of the bulb.

After the bulb has been formed as explained the entire end is served with spun yarn, or, for a better job, with fine copper wire. The socket, which was slipped over the rope before the bulb is completed, is now slipped into place on the bulb to finish the work.

Fig. 28: *A Back Splice in Wire Rope.* A seizing is clapped on the rope at about 12 to 18 inches from the end, the strands are unlaid and the fiber rope core is cut off short at the seizing. When this has been done proceed to tuck the strands over one and under two, against the lay, making four rounds of tucks in all.

Fig. 29: *A Short Splice in Wire Rope.* Clap on a seizing on the ends of each of the two ropes about two or three feet back from the ends. The length of the splice or rather the distance back at which the seizings are clapped on is of course determined

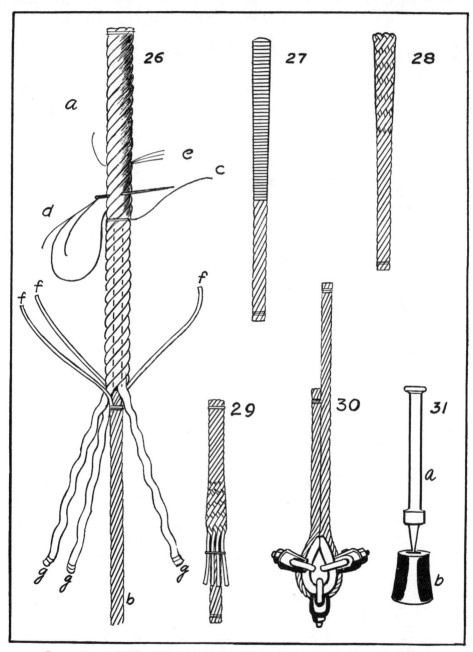

PLATE 14—A TAIL SPLICE AND A SHORT SPLICE IN WIRE ROPE

by the size of the rope, the larger the diameter the greater the length of the splice.

Unlay the ends of both ropes and open up the strands back to the seizings. Marry the two ropes and clap on a temporary seizing at the point where they join. At this point remove one of the temporary seizings clapped on before unlaying the strands. The splice must be clamped in a vise, otherwise it will be loose and might pull out.

Now select any strand on the side from which the seizing was removed and begin to tuck over one and under two strands against the lay of the rope. Make four rounds of tucks and then divide the strands into two parts. Bend one-half of each strand back and tuck each half-strand twice. This will make six rounds of tucks. The wire is then turned around and the seizing removed, after which the strands are spliced as was done with the other end of the rope. Pull up each strand as tightly as possible after each tuck. Remove the temporary seizings, beat out the splice well, working from the center to the ends, after which cut off the strands to finish the splice.

FIG. 30: *Turning in a Thimble*. When no vise is available a good method to use in clamping a thimble in the bight preparatory to forming an eye is to use ordinary wire clamps to hold the wire in place against the sides of the thimble. The clamps are removed after the splices have been made.

FIG. 31: *A Grommet Die* such as that shown is used for setting circular metal Grommets in canvas. The male half of the Grommet is placed on the die *b* and the female part of the Grommet is placed over the male half. The point of the punch *a* is placed in the hole in the Grommet and the punch is struck with a hammer until the Grommet is set up firmly in place.

Plate 15—A Long Splice, and Stitching Rawhide on Wire Rope

FIGS. 32A to F: These show a method employed to form a long splice in wire rope, differing slightly from the instructions on PLATE 10, FIGS. 17A, B and C. Mark off the length required for splicing (an equal distance on each end of the wire), which depends on the size of the rope. Proceed by unlaying three alternate strands from each end to within six inches of the mark on the left side of FIG. 32A.

Continue by cutting off the three remaining strands and also the core; then bend back the ends of the strands that were cut off, as shown on the right side of FIG. 32A. The ends are now locked together snugly and closely so that each alternate long strand will pass respectively to the right and left, as shown in FIG. 32B.

The rope is now gripped on one side of the lock with tongs; one short strand is then unlaid and the corresponding long strand from the opposite side is laid up in its place to within four feet of the end. To simplify operations, the short strand is now cut off four feet from the cross; the two ends will measure eight feet apart. Lay up the other two long strands in the same way; the crosses will be eight feet apart, as shown in FIG. 32C.

The next operation is to work the ends in the body of the rope as a substitute for the core, thereby maintaining as nearly the same circumference in the rope as possible. Before starting this operation, serve the ends of each strand to give a firmer grip on them by the outside strands of the rope.

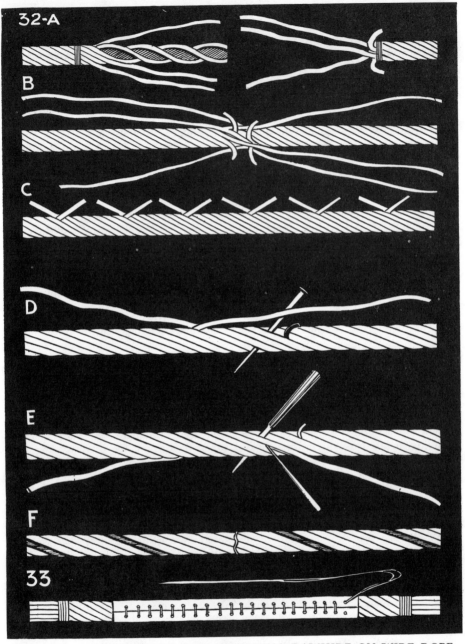

Plate 15—A LONG SPLICE, AND STITCHING RAWHIDE ON WIRE ROPE

Next grip the rope with tongs at the intersection where the two alternate strands meet in Fig. 32D. A marline spike is then driven through the center of the rope three strands in front of the strand to be worked in, and the core is hauled out. Run the spike back in the direction of the tongs; then hook a spoon under the end in the same place as the needle, as shown in Fig. 32E.

Two or three turns are now taken out of the strand after which it is twisted around the rope in the manner indicated. The handles of the needle and spoon are brought together at the same time, as illustrated. This operation forces the strand into the center of the rope; by twisting the needle around with the lay, the strand then will enter the rope from one side, at the same time forcing the core out on the opposite side. The core is cut off close to the end of the strand.

The opposite end is now worked in the rope in exactly the same manner and the other five pairs of strands are handled in the same way. Take care that the core is not run too far back, as this will tend to create flat places in the rope.

With ordinary Lay Rope, the ends of the strands are run towards the left of each other, as shown in Fig. 32F; with Lang Lay Rope the ends of the strands are run the opposite way or towards the right of each other.

Fig. 33: The accompanying illustration shows the proper method employed for stitching rawhide on wire rope. However, elkhide is preferable, as rawhide has a tendency to absorb salt and moisture and, consequently, to rust the wire.

Plate 16—Miscellaneous Wire Work

Figs. 1A, B and C: *Single Strand Liverpool Eye Splice.* Splicing single strand 1 x 19 stainless steel or aerial wire is usually looked upon as one of the most difficult of wire splices. But a rather simple and extremely efficient method of handling this type of wire follows:

A 1 x 19 wire, as shown in the illustration, consists of a heart of one wire, around which are six more wires, laid right-handed; around these, laid left-handed, are twelve more wires, making nineteen in all. To begin, the wire is first seized to a thimble, as shown; then two outer wires are selected and one inner—1, 2 and 3. Repeat this operation four times and you now will have five groups containing three wires each, making fifteen in all.

There now remain three wires and a heart; these are taken as the sixth group. Each group of wires are now laid up to form a single wire containing three strands, with the exception of the sixth group which contains four strands, as shown in illustration B. The next step is to begin the tucking which is done in precisely the same manner as the Liverpool Eye, Plate 8, appearing earlier in this book. The only difference in this splice is that everything is doubled; in other words, on the first tuck go under four wires instead of two, and on the second and all succeeding tucks go under two wires. The splice is continued in the spiral fashion of the Liverpool Eye and tapered down. After the splice has been completed, it is served with wire. It will test to 90% or better, depending on the quality of the splice. The first round of tucks is shown in C.

Fig. 2: *A Bar Seizing.* Lay one end of the seizing wire in the groove between two strands, as illustrated; then wrap the long end back over this portion. If a round bar is used, as shown here, the required tension in the wire is obtained by giving the free

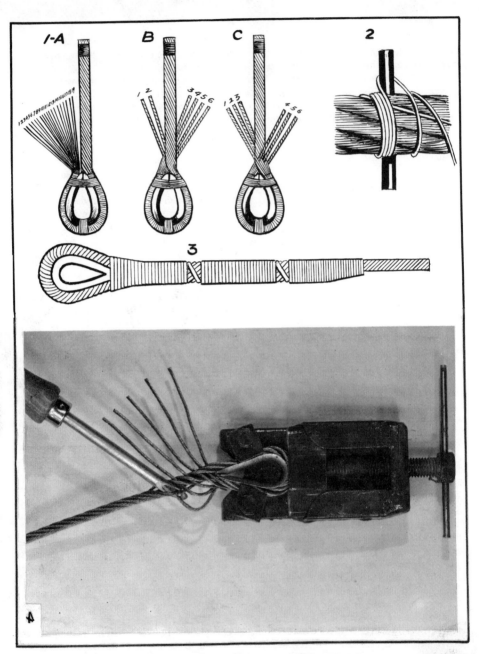

PLATE 16—MISCELLANEOUS WIRE WORK

end one or more turns around the rope. To finish off the ends of the seizing, the ends are twisted together in the usual manner.

FIG. 3: *A Served and Soldered Splice.* For instructions on the preparation of this splice, see the text on How to Splice a Served and Soldered Connection which appears later in this book.

FIG. 4: *Home-Made Rigger's Vise and Splicing Awl.* This vise is made from a piece of mild steel flat plate 2¼ inches wide, ½ inch thick and 5 inches long. On each side of this, a piece of angle 1¼ inches x ⅝ inch x 1/16 inch x 2¼ inches long is welded, one end of the ⅝-inch side being rounded. On the right-hand side is welded a block 1¼ inches deep, ½ inch wide and

2¼ inches long, with a ½-inch standard threaded hole through its center. Through this hole, put a ½-inch bolt 3 inches long with a 3/16-inch pin through its head. Another sliding block is made 2¼ inches long, 1 inch wide and ½ inch deep. This should have a ¼-inch V groove cut in its face. Next, two ¾-inch square blocks—each with a ¼-inch V groove cut in on one side and a 3/16-inch hole drilled through it—are bolted to the plate to complete the vise. The awl, seen on the left-hand side of the photograph, is made from a piece of 5/16-inch round rod, 6 inches long with a 3/16-inch hole drilled 4 inches deep. The rod is driven into a file handle to fit tightly; the end is then ground down to a beveled point. Its use in tucking needs no further explanation.

Plate 17—A Yachtsman's Tail Splice

FIGS. 1A to G: *A Yachtsman's Tail Splice* that is widely used on sailing yachts for the main halyards, backstay tail ropes, jib-sheets and, in many cases (such as with racing sloops), for the tail ropes on jib halyards. Although practically a necessity on most yachts, surprisingly few yachtsmen know how one should be made. A tail splice is the term used for joining a piece of wire rope to a piece of Manila hemp cotton or linen rope, presenting no noticeable bulk in the line at the point where the splice occurs, and permitting the rope to move freely through blocks and smoothly over sheaves. Rope is made up in this manner to haul in a jib or haul up a sail by hand with the rope (which can not be done when this portion is wire); and, when the sail is almost home, to take several turns around the winch with the wire portion and sweat it into place. The wire rope portion should be measured from the point where the sail is up tight to several feet beyond the winch, although this will vary, depending on the

size of the boat and the sails on which they are to be used. In any case, the wire should be long enough so that several turns can be taken around the winch when the point of the rope has been reached where the greatest strain must be applied. When racing in heavy blows, terrific strains are on the sails; if the wire is too short and the rope bears on the winch, it is liable to carry away—which, in most cases, means new wire, new rope and a new splice.

In the accompanying illustration, a piece of ¼-inch wire rope is spliced to a piece of ½-inch Manila rope. The Manila rope is shown at A and the wire rope at B.

To begin the splice, the wire rope is unlaid to a point 12 inches from the end. Three strands are seized at this point C, after which three of its alternate strands are unlaid another 12 inches where a seizing is put on at D. The next step is to unlay the Manila rope back 18 inches and put a light seizing on E, as shown in FIG. 1B. Marry the first three strands of the wire rope at C to

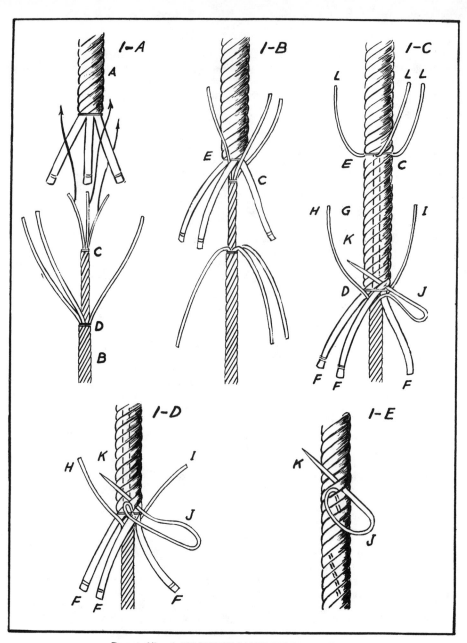

PLATE 17—A YACHTSMAN'S TAIL SPLICE

the Manila rope at E. They will then appear as shown in FIG. 1B.

Next, take the three strands of Manila rope F and lay them around the wire rope from C to D. They will then appear as in FIG. 1C. There will now be a piece of rope 12 inches in length, running from C to D with a wire core or heart, as shown by the dotted lines at G. Each one of the three strands of Manila rope (F) now are placed so that each falls between two strands of the wire H, I, J; a light seizing then is put on the Manila rope at D.

For the next step it is necessary to have a special needle. This can be made from a knitting needle by cutting it to about 4 inches in length and drilling a hole 1 inch deep, of sufficient diameter to hold the strand of wire snugly. The diameter of the hole will vary in accordance with the size of wire rope used.

The needle K is now forced through under the first strand, as shown in FIG. 1C; the wire strand is drawn through and pulled up tight. It will then appear as shown in FIG. 1D. The needle is now forced through the next strand of Manila as shown and again pulled up tight. This is continued, going around the rope until as much of the wire strand has been sewn into the rope as possible, as shown in FIG. 1E.

Next, strand I is sewn in a similar fashion, followed by H. This will complete the lower portion of the splice. The other three strands of wire L are now sewn into the rope in a similar manner. After all the strands are sewn in, they are cut off as closely as possible to the side of the rope and a whipping is put on over the ends as shown at M and N in FIG. 1F, PLATE 18. The Manila rope strands F are now unlaid and combed out and the yarns scraped down with a knife until they taper to a fine edge, 6 inches long. These are placed evenly and smoothly around the wire B and tied securely at ½-inch intervals with fine sail twine. Begin to serve or wind a length of marline around the wire beginning at O, which is 1 inch beyond the combed out Manila yarns. The use of a serving mallet at this point will facilitate the job. The serving is made 8 inches long. This will carry the serving 1 inch beyond the point where the Manila rope itself begins at D.

The whippings and servings are now rubbed well with beeswax and the job is complete. By following these instructions, you will have a beautiful and sturdy splice which will add much to the appearance and handling of a yacht. See FIG. 1G, PLATE 18. Also see PLATE 16 for a handy tool to use while preparing a splice of this kind.

Plate 18—A Yachtsman's Tail and Roll Splice

FIGS. 1F to G: *The Two Last Steps in Preparing a Yachtsman's Tail Splice,* as previously explained in PLATE 17 and its accompanying text.

FIGS. 2A to D: *A Yachtsman's Roll Splice* —a modified form of Flemish Eye in wire rope. Unlay the six strands from around the core down to the required length for the size of eye desired, at the same time taking care to keep them separated in two groups of three strands each without disturbing the lay of rope (*see* FIG. A).

After this has been done, cut the core out; proceed by taking one group of the two parts of three strands in each hand. Then, after determining the right allowance to use for working length of the strands, wind them together in the form of an Overhand Knot. This must be done so that the two groups of strands fall snugly into place with the lay of the rope which becomes apparent at once, since there is only one right way for the strands to fit together properly. The size of the eye can

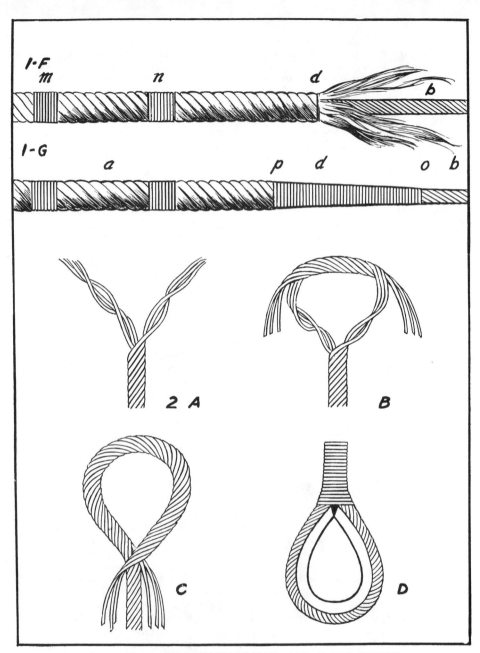

PLATE 18—A YACHTSMAN'S TAIL AND ROLL SPLICE

next be adjusted by pulling the two ends up accordingly (*see* FIG. B).

The work is now continued by winding each group of strands around, and in the groove of the lay of the opposite set of strands until the eye is neatly laid up down to the neck. This brings both ends lying along the standing part of the rope from the neck of the eye down. The ends are next unlaid as shown in FIG. C, then seized to the standing part. Do not tuck the strands for this will tend to weaken the rope and is unnecessary to prevent slipping.

The seizing should be applied with soft seizing wire, sweated on with a serving mallet, after the ends have first been distributed evenly around the body of the rope to insure a neat looking job (*see* FIG. D). The job can be soldered to finish off or a serving of marline can be applied if desired.

This form of eye splice requires far less time to finish than the ordinary methods of tucking the strands and is nearly 100% strong, whereas the tucking method is only 85% strong.

Chicago or Nash Splice

The following are instructions for making a tied splice in eight-strand regular lay rope with a hemp core.

Four alternate strands are unlaid from each end of the rope in the same manner as for six-strand rope, as explained in PLATE 10, FIG. 17. Proceed by following the same method until all four strands from the right side of the rope are laid up with the four corresponding strands from the left side of the rope. There are now four intersections, at which the projecting ends of the rope meet, each of which should be about 12 inches long. Place servings at an equal distance on each side of the points where each set of ends meet. Space the servings about one rope-lay apart.

Proceed by unlaying one-half of each strand and tie an Overhand Knot with them. Care should be used in choosing the proper half from each strand, so it will fill in properly with the lay of the rope. This knot is now pulled up snug with the use of clamps and a serving is placed around the

knot to keep it from slipping. With large rope, blocks and tackles are often necessary to hold the knot tight while being served.

The same method is repeated for all four sets of the eight-strands. There are now four ends of half-strands, from each set of two-strands, to be tucked in the rope. The serving is now removed from the Overhand Knots. Tuck the ends of each half-strand by driving a marlinespike under each two strands of the rope that lay next to the half-strands; insert the half-strands in a direction at right angles to the strands in the rope.

The operation is finished by using a wooden mallet to hammer the rope into proper shape at the points where the strands have entered, driving the ends of the projecting wire into the grooves between the strands in the rope. When using plow steel or blue center wire rope the ends of the strands should be cut off very close because this type of wire is very stiff and the ends of the strands are difficult to manage.

How to Splice Aircraft Cord

The following series of operations covers complete details of the *Official U.S. Navy Specification PS-2 for splicing aircraft cord*.

When splicing aircraft cord, the core, whether of cotton or wire strand, is always cut off close to the thimble.

NOMENCLATURE: The following definitions of splicing terms are applicable to this specification:

Standing wire: The measured length of cable which will be used when the splice is completed.

Bight: The loop of cable passing around the thimble.

Free end: The loose end which is unstranded for splicing.

POUNDING: All pounding of the splice should be done with a small mallet made of wood, fiber, copper or rawhide faced metal. The anvil on which the splice is pounded should be made of hardwood.

FORMING THE TERMINALS: The cable should be snugly bent around the proper size of thimble and clamped, the points of the thimble having been previously bent back at an angle of approximately 45 degrees to permit a close splice. If a regular splicing clamp is not available, the cable may then be held in place in the thimble by binding with cord. The length of the free end of the cable should be from 2 to 3 inches longer than required to produce the necessary number of tucks.

FIRST OPERATION: Choose the free strand lying directly under the thimble points and separate it from the rest of the free ends by inserting the marlinespike under it and spiralling the spike around the free end.

SECOND OPERATION: Insert the marlinespike under the first three strands of standing cable, to the right of the thimble points. Insert the loose free strand under the three standing strands, tucking from left to right.

THIRD OPERATION: Choose the second free strand to the left of the thimble points and separate from others as before described. This wire should be the next to the left of the first free strand tucked. Insert the marlinespike under first two standing strands to the right of the thimble points and tuck the loose end under these two standing strands, tucking from left to right.

FOURTH OPERATION: Choose the third free strand to the left of the thimble points and separate as before. Insert the marlinespike under one standing strand to the right of the thimble points and tuck the loose free end under this standing strand, tucking from left to right.

FIFTH OPERATION: Choose the center or core wire and separate it from the other free ends. It can be distinguished easily from the outside wires by its smooth, even surface. The outside wires are waved evenly by being coiled around the center or core wire. Tuck the center free end under the last two standing strands used, bringing it out in the same opening as the second free end which was tucked. The core wire should lie between the vise and the second free end.

SIXTH OPERATION: Choose the free strand which lies last to the left of the thimble points and separate as before. Insert the marlinespike from right to left under the two standing strands immediately to the left of the thimble points. Insert the loose free end from right to left under these two standing strands. This is the only place in the first round of tucks where the tuck is made from right to left.

To complete the sixth operation, pull the free end of the last strand tucked, until the loop disappears and there is no slack in the strand.

SEVENTH OPERATION: Separate the two remaining free ends. Insert the marlinespike under the first standing strand to the left of the thimble points. Choose the free end which lies next, in order of rotation, to the first three wires tucked. This should be the fourth free strand to the left of the thimble points. Tuck this free strand under the lifted standing strand, tucking from left to right.

EIGHTH OPERATION: With the marlinespike, lift the second standing strand to the left of the thimble points. Tuck the remaining free strand under the lifted standing strand. Tuck from left to right.

NINTH OPERATION: The finished condition of a splice depends considerably on this operation which should be observed very closely. Choose the tucking strand which lies in the opening with the center strand. Grasp it firmly in a pair of fairly heavy pliers and pull sharply upon it several times, away from the thimble. Then, keeping the strand always taut, describe a half circle with the pliers and pull steadily back into the throat of the thimble. (The throat of the thimble is the joint of the two ends lying between the four points, previously turned back at a 45 degree angle.) Continue this operation completely around the standing cable until each of the tucking strands has been pulled down away from the thimble and then back against the throat. Next pull down the free center strand, but do not pull it back. Having finished this, you are ready for the next operation.

TENTH OPERATION: Choose the tucking strand to the right of the center free strand and tuck it over one standing strand and

under the next standing strand. Tuck from right to left. Continue this operation around the standing cable to the left until you have tucked all of the six tucking strands. Do not tuck the free center strand on this or any succeeding series of tucks. After tucking all six tucking strands, repeat the NINTH OPERATION. This constitutes the second series of tucks.

ELEVENTH OPERATION: The third series of tucks is made exactly the same as the second series. (Follow with the NINTH OPERATION.)

TWELFTH OPERATION: Cable which is less than $1/4$ inch in-diameter is now ready for tapering; cable $1/4$ inch and over is given another or fourth tuck, using the full strands and proceeding as described for the eleventh operation. For the purpose of tapering, each of the six strands should have its cross section reduced one-third by fraying out one-third of the wires. The first tapering tuck is then made with the six $2/3$ strands in the same manner as that described for the full strands. (Repeat the NINTH OPERATION.)

THIRTEENTH OPERATION: The core wire or free center strand is now cut off as close as possible to the splice. The six $2/3$ strands are reduced one-half in cross section by fraying out one-half of the wires, making them one-third the size of the original strand. A series of tucks should now be made using the six $1/3$ strands. (Repeat the NINTH OPERATION.) This completes the tucking of the strands.

POUNDING THE SPLICE: Bend the thimble points back into position and tap them snugly against the splice. Using a hardwood anvil and a hardwood or rawhide faced mallet, start at the thimble points and drive the tucking strands from right to left just as they were tucked. Roll the splice on the anvil while pounding in order that the strands may receive an even tautness. When the splice has been made taut and hard, cut off the tucking strands as close as practicable. Pound the taper slightly to lay the sheared ends.

SERVING THE SPLICE: Using the proper size of cord, start about ¼ inch below the end of the taper and wrap smoothly up the taper until all of the sheared ends are covered. Then take five or six loose coils over thumb or finger and insert the end of the serving cord through the loose coils from the thimble and toward the taper. Now wrap the loose coils firmly over the inserted end and pull up any slack by drawing the inserted end down toward the taper. Cut off close to the serving and beat it lightly with the mallet to smooth it down. Apply two coats of shellac to the serving in order to make it as nearly waterproof as possible.

U.S. Army-Navy Aeronautical Splice

The following series of operations cover complete details of the *Official U.S. Army-Navy Aeronautical Specification AN-S-43 for splicing flexible aircraft cable.*

Flexible Cable: The cable should be cut to length by mechanical means only, such as a fine tooth hacksaw. The use of oxyacetylene torches in any manner should not be permitted.

OPERATIONS FOR FLEXIBLE CABLE: The procedure for the manufacture of terminal splices for flexible cable is described in the following paragraphs. The splice should have five full tucks. For 1/16-inch 7 x 7 flexible steel cable, the procedure described later for non-flexible cable may be used as an optional method.

FORMING OF TERMINALS: The cable is bent snugly around the proper size of thimble and clamped, the points of the thimble having previously been bent back at an angle of approximately 45 degrees to permit a close splice. If a regular splicing clamp is not available, the cable may be held in place on the thimble by binding with cord.

The length of the free end of the cable should be about 8 inches for sizes up to and including 5/32-inch 7 x 19 cable. For cable of larger sizes, the length should be slightly greater. The standing wire should be clamped in a vise with the free end to the left of the standing wire and away from the operator.

The same procedure is followed from the first to the eleventh operation as for the official U.S. Navy method that has been previously described. The splicing is now continued with the following operations.

TWELFTH OPERATION: The six tucking strands are now split in half, and a series of tucks are made in the same manner as that described for the full strands, but using only one-half of each strand. (Follow with the NINTH OPERATION, described in How to Splice Aircraft Cord.)

THIRTEENTH OPERATION: The core wire or free center strand should now be cut off as close as possible to the standing cable. The six ½ strands, which were tucked in the TWELFTH OPERATION, are again halved, making them one-fourth the size of a full strand. A series of tucks is now made using six ¼ strands. (Repeat the NINTH OPERATION.) This completes the tucking of the strands. The splice is now pounded and served in the same manner as for the Navy method previously explained.

How to Splice a Served and Soldered Connection

The following series of operations cover complete details of the *Official Army-Navy Aeronautical Specification AN-S-43 for splicing a served and soldered connection,* which is frequently used for several of the smallest sizes of aircraft cord.

Non-flexible cable: Before cutting the cable the wires must be soldered or welded together to prevent slipping. The preferable process is to thoroughly tin and solder the cable for 2 or 3 inches by placing in a solder trough and finishing smooth with a soldering tool. The cable may be cut diagonally to conform to the required taper finish. The cable should be cut to length by mechanical means, such as a fine tooth hacksaw.

OPERATIONS FOR NON-FLEXIBLE CABLE: The procedure for the manufacture of terminal splices for non-flexible cable is described in the following paragraphs. This terminal must not be used on flexible cable, except that it may be used for 1/16-inch 7 x 7 flexible cable.

FORMING OF SPLICE: After soldering and cutting, the cable is securely bent around the proper size thimble and clamped, taking care that the cables lie close and flat, and that the taper end for finish lies on the outside. If it is necessary to trim the taper at this point in the process, it is preferable that it be done by nipping; grinding will be permitted, provided a steel guard at least 3 inches long and 1/32 inch thick is placed between the taper end and the main cable during the operation, and provided the heat generated from the grinding does not melt the solder and loosen the wires.

SERVING: Serving may be done by hand or machine; in either case, each serving convolution must touch the adjoining one and be pulled tightly against the cable, with spaces for permitting a free flow of solder and inspection, as shown in FIG. 3, PLATE 16. Serving wire should follow specifications for terminal dimensions in all details. The table of specifications will be found at the end of this explanation.

SOLDERING: Care must be exercised to prevent drawing of the temper of any cable wires by excessive temperatures or duration of applied heat. The flux used in this soldering should be a compound of stearic acid and rosin of the following proportions: stearic acid, 25 to 50%, rosin, 50 to 75%. Salammoniac, or other compound having a corrosive effect, must not be employed, either as a flux or for cleaning the soldering tools.

Soldering is accomplished by immersing the terminal alternately in the flux and in the solder bath, repeating the operation until thorough tinning and filling with solder under the serving wire and thimble is obtained. The temperature of the solder bath and place where the terminal is withdrawn should not be above 450 degrees F. A soldering iron may be used in the final operation to give a secure and good appearing terminal. Care must be taken that the solder completely fills the space under the serving wire and thimble. A slightly hollowed cast-iron block to support the splice during soldering may help in securing best results. Abrasive wheels or files for removing excess solder must not be used.

ALTERNATE PROCESS FOR NON-FLEXIBLE CABLE: As an alternate process of making terminals for non-flexible cable, the oxyacetylene cutting method and the pre-soldering method (soldering before wrapping) are permissible only under the following conditions: (1) the process of cutting securely welds all wires together; (2) the annealing of the cable does not extend more than one cable diameter from the end; (3) no filing has been done either before or after soldering; (4) for protection during the operation of grinding the tapered end of the cable, a steel guard at least 3 inches in length and 1/32 inch thick will be placed between the taper and the main cable; (5) the heat from grinding will not draw the temper of the cable.

The following table gives the proper size of cable and serving wire to use with the served and soldered splice:

DIAMETER OF CABLE	TERMINAL DIMENSIONS			
	L ± 1/8	D ± 1/16	C ± 1/16	SERVING WIRE B & S GAUGE
1/16	2	9/16	1/8	24
3/32	2-1/2	3/4	1/8	24
1/8	2-7/8	7/8	1/8	24
5/32	3-1/4	1	1/8	24
3/16	3-5/8	1-1/8	1/8	20
7/32	4	1-1/4	1/8	20
1/4	4-1/2	1-3/8	3/16	20
5/16	5-1/4	1-5/8	3/16	20
3/8	6-1/4	1-15/16	1/4	18
7/16	7	2-3/16	1/4	18
1/2	8	2-1/2	1/4	18

Wire Rope Sizes and Characteristics

GALVANIZED IRON RIGGING AND GUY ROPE

Composed of 6 Strands and a Hemp Center, 7 Wires to the Strand

Diameter in inches	Approx. circumference in inches	Approx. weight per foot	Breaking strength in tons of 2000 lbs.	Circum. of manila rope of nearest strength
1¾	5½	4.60	37.00	10
1⅝	5⅛	3.96	32.40	9
1½	4¾	3.38	27.70	8½
1⅜	4⅜	2.84	23.70	7½
1¼	3⅞	2.34	19.90	7
1⅛	3½	1.90	16.50	6
1 1/16	3⅜	1.70	14.80	5½
1	3⅛	1.50	13.20	5¼
⅞	2¾	1.15	10.20	4¾
¾	2⅜	.84	7.10	3¾
⅝	2	.59	5.30	3¼
9/16	1¾	.48	4.32	3
½	1⅝	.38	3.43	2½
7/16	1⅜	.29	2.64	2¼
⅜	1⅛	.21	1.95	2
5/16	1	.15	1.36	1½
9/32	⅞	.125	1.20	1⅜
¼	¾	.090	.99	1¼
7/32	11/16	.063	.79	1⅛
3/16	⅝	.040	.61	1

EXTRA PLIABLE HOISTING ROPE

Composed of 6 Strands and a Hemp Center, 37 Wires to the Strand

Diameter in inches	Approx. circumference in inches	Approx. weight per foot	Breaking strength in tons of 2000 lbs.
3½	11	19.00	451.0
3¼	10¼	16.37	392.0
3	9⅜	13.95	337.0
2¾	8⅝	11.72	285.0
2½	7⅞	9.69	237.0
2¼	7⅛	7.85	194.0
2	6¼	6.20	155.0
1¾	5½	4.75	119.5
1⅝	5⅛	4.09	103.3
1½	4¾	3.49	88.2
1⅜	4⅜	2.93	74.3
1¼	3⅞	2.42	61.5
1⅛	3½	1.96	49.9
1	3⅛	1.55	39.5
⅞	2¾	1.19	30.5
¾	2⅜	.87	22.8
⅝	2	.61	16.1
½	1⅝	.39	10.6
⅜	1⅛	.22	6.1
¼	¾	.10	2.8

GALVANIZED STEEL MOORING LINES AND HAWSERS

Composed of 6 Strands and a Hemp Center, each Strand composed of 24 Wires and a Hemp Core

Diameter in inches	Approximate circumference in inches	Approximate weight per foot	Breaking strength in tons of 2000 lbs.	
			Plow steel	Cast steel
2 1/16	6½	5.87	118.00	98.00
2	6¼	5.52	112.00	92.00
1 13/16	5¾	4.53	92.30	76.20
1¾	5½	4.23	86.20	71.20
1⅝	5⅛	3.64	74.50	61.60
1½	4¾	3.11	63.60	52.60
1⅜	4⅜	2.61	53.60	44.40
1¼	3⅞	2.16	44.40	36.70
1⅛	3½	1.75	36.00	29.90
1	3⅛	1.38	28.50	23.70
⅞	2¾	1.06	22.00	18.30
¾	2⅜	.78	16.40	13.60
⅝	2	.54	11.60	9.59
½	1⅝	.35	7.63	6.37
⅜	1⅛	.194	4.40	3.67

GALVANIZED STEEL HAWSERS

Composed of 6 Strands and a Hemp Center, 37 Wires to the Strand

Diameter in inches	Approximate circumference in inches	Approximate weight per foot	Breaking strength in tons of 2000 lbs.
2⅜	7½	8.74	173.3
2¼	7⅛	7.85	156.2
2⅛	6⅝	7.00	140.2
2	6¼	6.20	125.0
1¾	5½	4.75	96.5
1⅝	5⅛	4.09	83.4
1½	4¾	3.49	71.2
1⅜	4⅜	2.93	60.0
1¼	3⅞	2.42	49.7
1⅛	3½	1.96	40.3
1 1/16	3⅜	1.75	36.0
1	3⅛	1.55	31.9
⅞	2¾	1.19	24.6
13/16	2½	1.02	21.3
¾	2⅜	.87	18.3

Rope Splicing

Splicing has been dealt with heretofore in many different ways, but it has never been covered thoroughly until now. Like most of the other types of rope work it seems to have reached a certain stage and then never to have been developed or expanded any further.

Besides all the practical methods which are in common use there are presented in this section many universal examples, such as the Eight-Strand Braided Eye or Signal Halyard Splice, the Grecian Splice and various types of Halter Splices. Many examples of splicing have been carried out or extended from three to four and then six strands for the purpose of giving additional knowledge in the art of being able to splice all kinds of rope regardless of its type or the number of its strands.

Eye Splicing, Short Splicing and Long Splicing by both the regular and sailmaker's methods are explained and illustrated fully, as are the many other different examples of both practical and unusual types.

Plate 19—Three- and Four-Strand Eye and Back Splices

Fig. 1: The *Three-Strand Inverted Wall Eye Splice* is begun by making the first tuck of a Three-Strand Eye Splice. Then an Inverted Wall Knot is tied around the standing part of the rope. The ends are spliced into the rope to finish off.

Fig. 2: The *Four-Strand Inverted Wall Eye Splice*.

Fig. 3: The *Three-Strand Double Manrope Eye Splice* begins with an Eye Splice tuck; then follows the regular Manrope Knot. This knot will lead out toward the eye on the last tuck; therefore the ends are cut off short, instead of being spliced in.

Fig. 4: The *Four-Strand Double Manrope Eye Splice*.

Figs. 5A, B, and C: The *Three-Strand Back Splice* starts with a Three-Strand Crown. Then tuck each strand over the next strand and under the second strand, and pull taut. Fig. 5A shows the first tuck started, with strand *c* going over the first strand and under the second. Fig. 5B illustrates the second tuck, with strand *b* tucked in the same manner as strand *c*. Strand *a* follows the same procedure. Make several tucks, and taper the splice down by splitting each strand in half for each remaining tuck. Then pull taut and trim all strands. Fig. 5C shows the splice as completed.

Figs. 6A and B: The *Four-Strand Back Splice* is made in the same way as the three-strand splice. Fig. 6A shows the Four-Strand Back Splice with the first set of tucks finished. Fig. 6B represents the completed splice.

Figs. 7A, B, and C: The *Three-Strand Sailmaker's Back Splice* begins with a Crown, as in the regular splice. But instead of going over one and under one, each tuck is made over, around, and then under the next or following strand, which brings them out with the lay of the rope. Fig. 7A shows strand *b* tucked in this fashion. Strands *c* and *a* are tucked in the same way, until they take on the appearance of Fig. 7B. Fig. 7C shows the splice as it looks when pulled taut, tapered, and finished. This neat form of splicing is used by sailmakers for awnings.

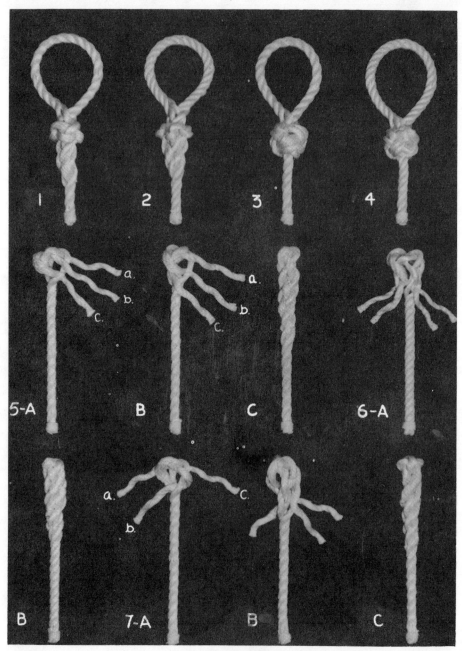

PLATE 19—THREE- AND FOUR-STRAND EYE AND BACK SPLICES

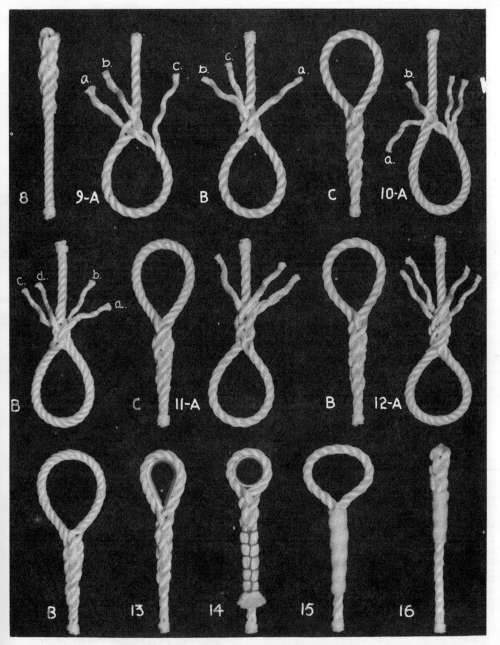

PLATE 20—MISCELLANEOUS EYE SPLICES

Plate 20—Miscellaneous Eye Splices

FIG. 8: The *Four-Strand Sailmaker's Back Splice.*

FIGS. 9A, B, and C: The *Three-Strand Eye Splice* is made as follows: unlay the rope a sufficient distance, and make an eye of the required size; tuck the bottom strand under one strand against the lay, and place the middle strand under the next strand in the same manner. Turn the splice over, and tuck the remaining strand as already described. Strands *a* and *b* are shown tucked in FIG. 9A. FIG. 9B shows the splice turned over, with strand *c* tucked. Tuck each strand a second time against the lay, over one and under one, and repeat as many more times as desired. After tapering down to finish off, the splice will appear as in FIG. 8C. This splice and all other splices can be rolled under the foot or hammered down to insure a neat, close-fitting job.

FIGS. 10A, B, and C: The *Four-Strand Eye Splice* is a repetition of the three-strand method, except at the beginning. The bottom strand is tucked under two strands, and the other strands are tucked under one each. Strand *a* in FIG. 10A is shown tucked under two strands as described, and strand *b* under one strand. Turn the splice over and tuck strands *c* and *d* accordingly. The splice will then appear as in FIG. 10B. Continue by tucking over one and under one against the lay, until the required number

of tucks have been made. FIG. 10C shows the splice as it looks when finished.

FIGS. 11A and B: The *Three-Strand Sailmaker's Eye Splice,* First Method, is begun by unlaying the rope and making the first tucks as though for the regular Eye Splice. Then continue by making the sailmaker's tucks as shown in FIG. 11A. When the required number have been made, taper down and cut the strands off short. FIG. 11B shows the splice as it appears when completed.

FIGS. 12A and B: The *Four-Strand Sailmaker's Eye Splice,* First Method, is made in the same way as the three-strand knot. FIG. 12A shows the splice with the first set of sailmaker's tucks completed. The finished knot is illustrated in FIG. 12B.

FIG. 13: The *Thimble Eye Splice* consists of a thimble spliced into an eye. It is used to prevent a hook from chafing the rope.

FIG. 14: The *Round Thimble Eye Splice with Ends Frayed* is a round thimble, spliced into an eye with the ends frayed out, and hitched along the body of the rope with Marline.

FIG. 15: The *Eye Splice Served* has the eye spliced in, tapered down, and served.

FIG. 16: The *Back Splice Served* represents a regular Back Splice tapered down and served in the usual manner.

Plate 21—Miscellaneous Eye Splices

FIG. 17: The *Three-Strand Wall Eye Splice* consists of a Wall Knot tied around the body of the rope, after splicing in the first round of tucks. The ends are spliced back into the rope, and cut off short.

FIG. 18: The *Four-Strand Wall Eye Splice.*

FIG. 19: The *Three-Strand Lanyard Eye Splice* has a Lanyard Knot tied around the body of the rope, after the first tucks have been spliced in. The knot is finished off by

splicing and trimming the ends, as shown in FIG. 17.

FIG. 20: The *Four-Strand Lanyard Eye Splice.*

FIG. 21: The *Three-Strand Turk's Head Weave Eye Splice* is based on the Turk's Head Weave.

FIG. 22: The *Four-Strand Turk's Head Weave Eye Splice.*

FIG. 23: The *Three-Strand Walled*

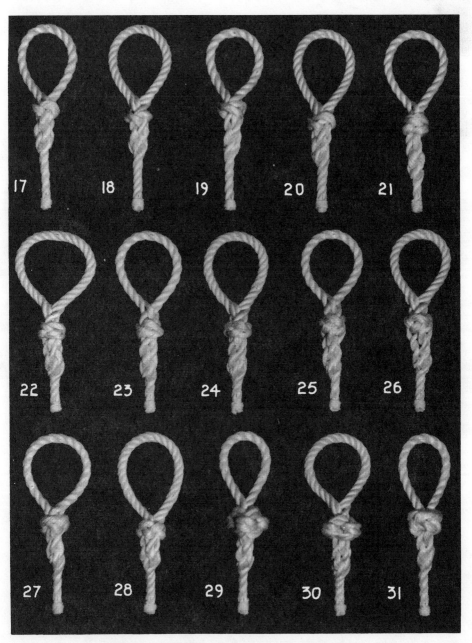

PLATE 21—MISCELLANEOUS EYE SPLICES

Crown Eye Splice utilizes the Walled Crown Knot.

Fig. 24: The *Four-Strand Walled Crown Eye Splice.*

Fig. 25: The *Three-Strand Single Stopper Eye Splice* is made by tying a Single Crown and Wall, and splicing in the strands.

Fig. 26: The *Four-Strand Single Stopper Eye Splice.*

Fig. 27: The *Three-Strand Single Diamond Eye Splice* begins with the eye spliced into the rope in the usual manner. Then form a Diamond Knot, and finish off in the same manner as in the preceding knots.

Fig. 28: The *Four-Strand Single Diamond Eye Splice.*

Fig. 29: The *Three-Strand Star Eye Splice* is formed with a Star Knot, and finished off as usual.

Fig. 30: The *Four-Strand Star Eye Splice.*

Fig. 31: The *Three-Strand Sennit Eye Splice* is begun by splicing the eye into the rope, just as in all knotted eye splices. Follow this with a Sennit Knot, and finish off by splicing and trimming the strands in the regular way.

Plate 22—Three- and Four-Strand Eye Splice Combinations

Fig. 32: The *Four-Strand Sennit Eye Splice.*

Fig. 33: The *Three-Strand Matthew Walker Eye Splice* is begun by splicing the first set of eye tucks. Then a Matthew Walker Knot is tied. The ends are spliced into the rope and cut off short to finish.

Fig. 34: The *Four-Strand Matthew Walker Eye Splice.*

Fig. 35: The *Three-Strand Outside Matthew Walker Eye Splice* is tied with the Outside Matthew Walker.

Fig. 36: The *Four-Strand Outside Matthew Walker Eye Splice.*

Fig. 37: The *Three-Strand Single Manrope Eye Splice* follows the Eye Splice with first a Wall, and then a Crown, which is pulled up and spliced back into the rope.

Fig. 38: The *Four-Strand Single Manrope Eye Splice.*

Fig. 39: The *Three-Strand Double Lanyard Eye Splice* uses the Lanyard Knot doubled, which is tied around the standing part of the rope in the usual way.

Fig. 40: The *Four-Strand Double Lanyard Eye Splice.*

Fig. 41: The *Three-Strand Double Diamond Eye Splice* is based on a Double Diamond Knot.

Fig. 42: The *Four-Strand Double Diamond Eye Splice.*

Fig. 43: The *Three-Strand Double Walled Crown Eye Splice* utilizes a Walled Crown Knot doubled.

Fig. 44: The *Four-Strand Double Walled Crown Eye Splice.*

Fig. 45: The *Three-Strand Double Stopper Eye Splice* has a Crown and Wall doubled in the usual manner to form a Stopper Knot in the Eye Splice.

Fig. 46: The *Four-Strand Double Stopper Eye Splice.*

Plate 23—Six-Strand Eye and Back Splices

Fig. 47: The *Six-Strand Back Splice* is made by first removing the heart, then forming the splice as in the three- and four-strand methods.

Fig. 48: The *Six-Strand Sailmaker's Back Splice* is tied in the same way as the three and four-strand versions.

Figs. 49A, B, and C: The *Six-Strand Eye*

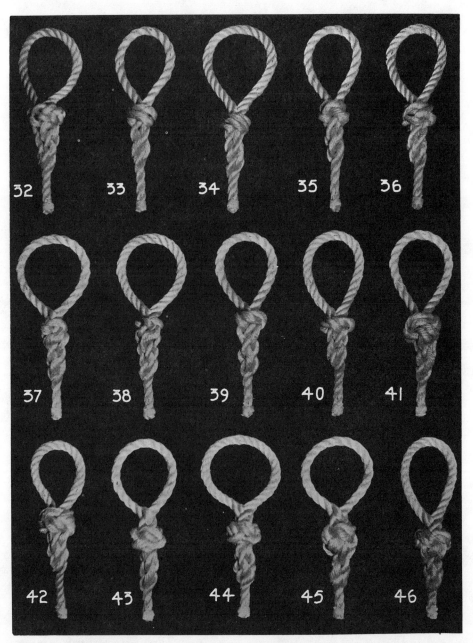

PLATE 22—THREE- AND FOUR-STRAND EYE SPLICE COMBINATIONS

Splice is begun by unlaying the strands the required amount and placing a whipping on to keep them from unlaying further. Next, remove the heart, and bend the rope in from right to left toward the standing part. Start with the underneath strand nearest the standing part of the rope, and tuck it through the middle, or under three strands, as shown with strand *a* in FIG. 49A. Then take the following strand on the right of the first tuck and tuck it through two strands, after first starting it through in the same place as the first tuck. But be careful to keep it slightly ahead, in order to bring it out in its proper place one strand above the first tuck, as shown with strand *b* in FIG. 49A. The next strand is tucked under one (*See* strand *c* in FIG. 49A), after which the Splice is turned over, and strands *d*, *e*, and *f* are then tucked under one strand each against the lay. It will then appear as in FIG. 49B. Make two or three

tucks over one and under one, then taper down and cut off. FIG. 49C shows the Splice as it will look when finished.

FIG. 50: The *Six-Strand Sailmaker's Eye Splice*, First Method, is made in the same manner as the three and four-strand methods, after first removing the heart, and starting the first tucks in the same manner as illustrated in FIG. 49.

FIG. 51: The *Six-Strand Double Manrope Eye Splice* is tied by first unlaying the strands the proper distance. Make the first set of tucks, then tie a Manrope Knot around the standing part of the rope. Pull taut, and cut off the strands to complete the knot. At times, the end of a rope that runs close to a block requires an eye; and as an ordinary Splice would have a tendency to jam, a knot of this sort is preferred. Any one of the various other Eye Splice Knots that are quick and easy to form might also be used.

Plate 24—Miscellaneous Six-Strand Eye Splices

FIG. 52: The *Six-Strand Wall Eye Splice* is begun, as usual, by unlaying the rope a sufficient distance. Next, seize it, remove the heart, and splice an eye into the rope as shown in PLATE 23, FIG. 49. Then make the first set of tucks, and tie the knot in the same manner as the three and four-strand methods. Splice the ends into the rope in the ordinary fashion to finish off.

FIG. 53: The *Six-Strand Lanyard Eye Splice*.

FIG. 54: The *Six-Strand Double Lanyard Eye Splice*.

FIG. 55: The *Six-Strand Single Diamond Eye Splice*.

FIG. 56: The *Six-Strand Double Diamond Eye Splice* follows the single-strand method, except that it is doubled.

FIG. 57: The *Six-Strand Single Stopper Eye Splice*.

FIG. 58: The *Six-Strand Double Stopper Eye Splice* is made in the same way as the single-strand knot, then doubled.

FIG. 59: The *Six-Strand Turk's Head Weave Eye Splice*.

Plate 25—Miscellaneous Six-Strand Eye Splices

FIG. 60: The *Six-Strand Star Eye Splice*.

FIG. 61: The *Six-Strand Sennit Eye Splice*.

FIG. 62: The *Six-Strand Matthew Walker Eye Splice*.

FIG. 63: The *Six-Strand Overhand Matthew Walker Eye Splice*.

FIG. 64: The *Six-Strand Single Walled*

Crown Eye Splice.

FIG. 65: The *Six-Strand Double Walled Crown Eye Splice*.

FIG. 66: The *Six-Strand Inverted Wall Eye Splice*.

FIG. 67: The *Six-Strand Single Manrope Eye Splice* contains a Wall and Crown tied and spliced in the usual manner.

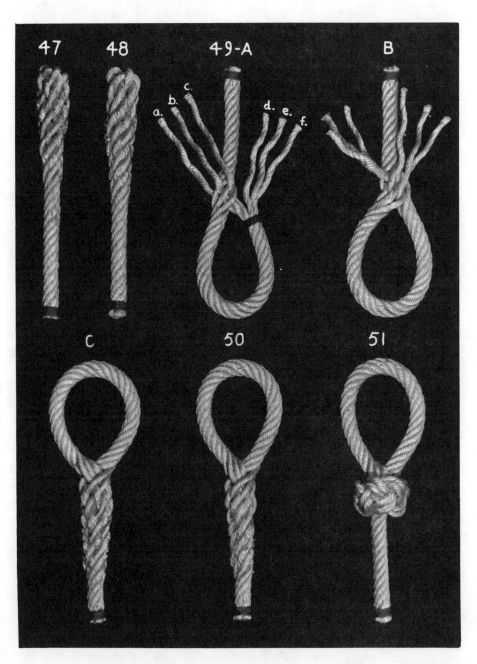

PLATE 23—SIX-STRAND EYE AND BACK SPLICES

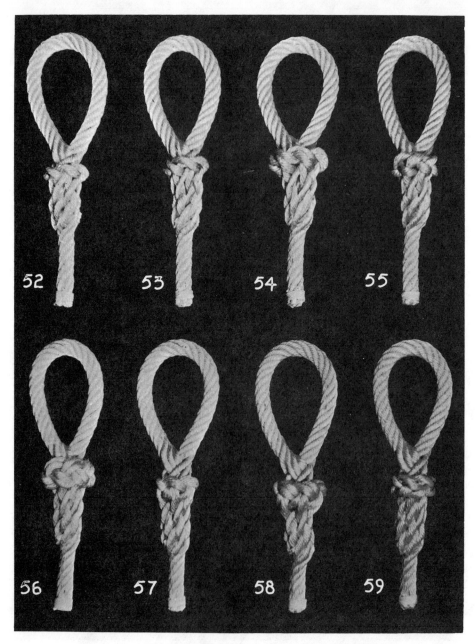

PLATE 24—MISCELLANEOUS SIX-STRAND EYE SPLICES

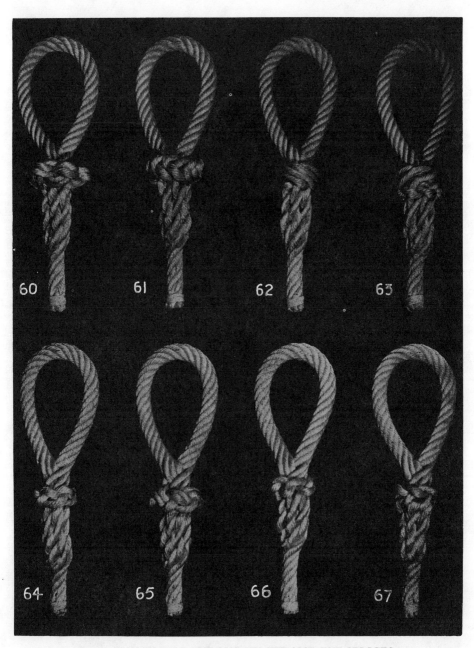

PLATE 25—MISCELLANEOUS SIX-STRAND EYE SPLICES

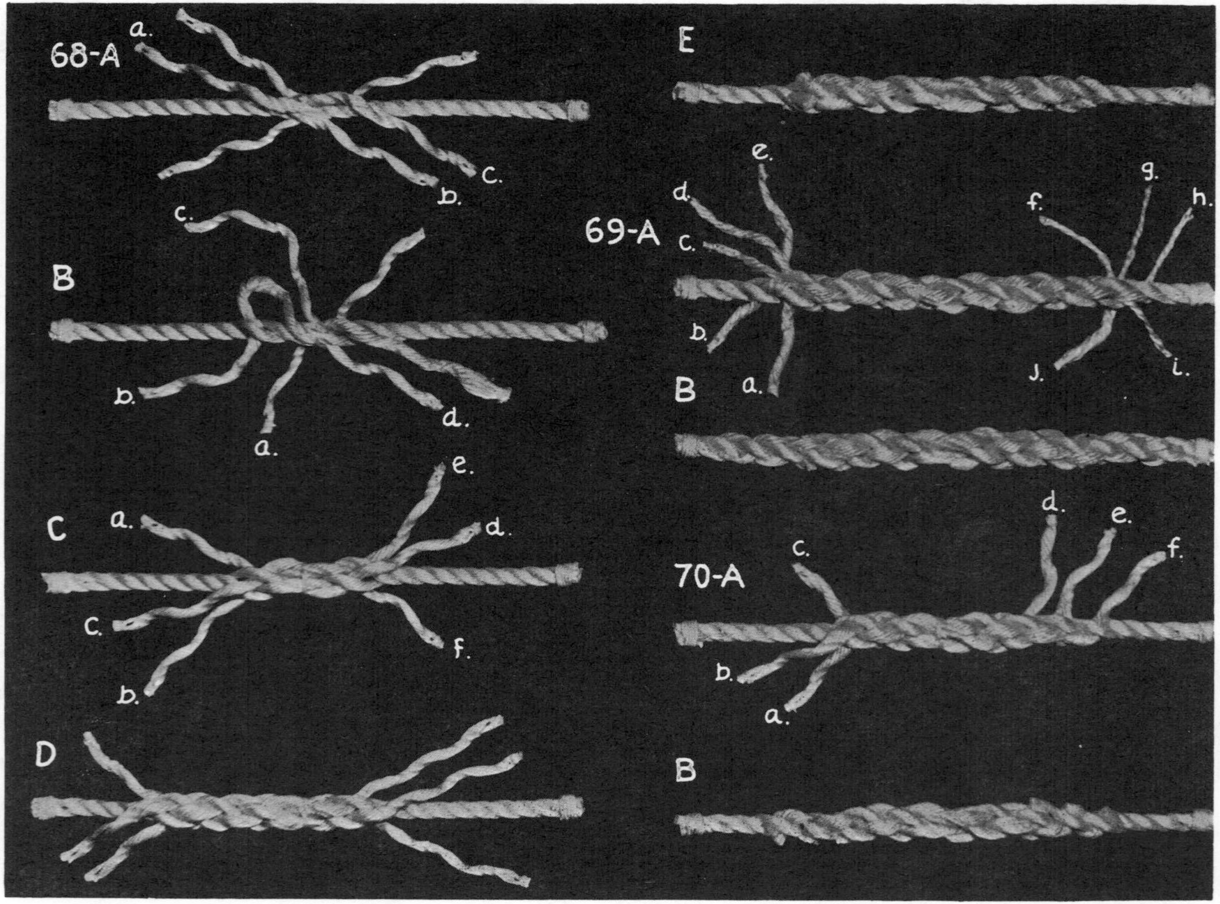

PLATE 26—THREE-STRAND SHORT SPLICES

Plate 26—Three-Strand Short Splices

Fig. 68A: The *Three-Strand Short Splice* is the strongest and most secure method of uniting two ropes. It is stronger than the Long Splice, but increases the diameter of the rope, so that the spliced portion of the rope may be unable to pass over the sheave of a block. To begin this Splice, unlay the ends of each rope, and marry the strands as shown in the illustration. In marrying a rope, one strand of one rope comes between two strands of the other, as shown by strand *a*, which comes between the two strands *b* and *c*.

B: The work will be found much simpler to execute if a temporary seizing is placed at the point where both ropes have been married. This will serve to hold the ropes in position until the first tucks are made. The strands on the right-hand side are, for the moment, let alone. Begin with any strand on the left. Tuck it over one and under one against the lay. Strand *b*, as shown in the illustration, is tucked over strand *d* and under the next strand. Either strand *a* or *c* may be tucked next.

C: After strands *a*, *b*, and *c* have been tucked, strands *d*, *e*, and *f* are tucked once each. The Splice is begun on both sides, there now being one round of tucks. Two more tucks are put in with each strand, making three rounds of tucks in all, which is the proper amount for a secure Splice.

D: This shows the Short Splice after three rounds of tucks have been made. The Splice is now rolled under the foot and pounded with a fid or mallet, to make it round and work the strands into place. The strands are then cut off, but not too close, or they may work out when tension is applied.

E: This illustration shows the finished Splice. It is good practice for the beginner to put a whipping on the end of each strand

to prevent unlaying. The ends can be sewn to the body of the rope with sail twine to make them more secure.

Fig. 69A: The *Short Splice Tapered,* First Method, is an especially neat form of Short Splice, because of its tapered strands. The tapering, however, should not be begun until three rounds of tucks have been made. Each strand is then split in half, as shown by strands *d* and *e*. In this case, *d* is tucked and *e* is cut off. Strand *b* has already been tucked, and its remaining half, *c*, is cut off. The full strand *a* has not yet been split. Two complete rounds of tucks are then made with each half-strand. They are halved again, making quarters, as shown by *f* and *g*. In this case, *f* is tucked and *g* is cut off. Strand *h* has already been tucked, and its other half *i* is to be cut off. The half-strand *j* has not yet been quartered. One round of tucks is made with the quartered strands.

B: After the Splice has been rolled and pounded, all the remaining strands are cut off, and the finished Splice will then appear as in the illustration. The Splice, it should be noted, will be found just as secure if only two rounds of full tucks are made before tapering.

Fig. 70A: The *Short Splice Tapered,* Second Method, is another and much more rapid method of tapering the Short Splice. First, marry the ropes and splice two rounds of tucks. Instead of tucking strand *a*, tuck both strands *b* and *c* once. Strands *a* and *b* are left as they are, and the remaining strand *c* is tucked again. When this has been done, the strands will appear in the position shown by strands *d*, *e*, and *f*, only pointing downward rather than up.

B: The Splice is then rolled and pounded and the strands cut off. The finished Splice is pictured in this illustration.

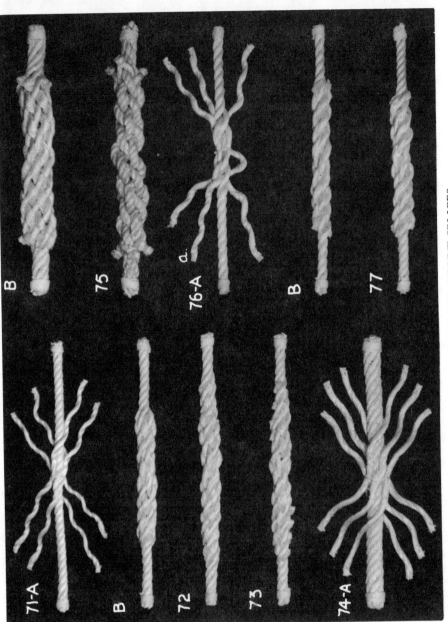

PLATE 27—FOUR-STRAND SHORT SPLICES

Plate 27—Four-Strand Short Splices

FIGS. 71A and B: The *Four-Strand Short Splice* has the strands married and spliced by means of the same method used for three-strand rope. FIG. 71B shows how the completed Splice will look after the required three rounds of tucks have been made and the strands cut off.

FIG. 72: The *Four-Strand Splice Tapered*, First Method, follows the procedure used for the Tapered Splice shown in PLATE 26, FIG. 69.

FIG. 73: The *Four-Strand Splice Tapered*, Second Method, is tied in the same way as the Tapered Splice in PLATE 26, FIG. 70.

FIGS. 74A and B: The *Six-Strand Short Splice* is begun by first placing a temporary whipping on each rope before unlaying the strands. Next, unlay the strands to the whipping, and remove the heart. Marry the strands in the same manner as the three and four-strand rope. And place a temporary whipping at the point where they join, to hold the ropes in place until the tucks have been made. After making one round of tucks, the temporary whippings on each rope are cut off. Then after two more rounds of tucks have been made, the whipping put on after marrying the strands is removed. The Splice is pounded and rolled under the foot, and the strands cut off short to finish. FIG. 74B shows the Splice completed.

FIG. 75: The *Short Splice (Cable-Laid Rope)* is made by first unlaying the three strands of each rope. Marry and splice them in the same manner as the three-strand rope. After three rounds of tucks have been made, the three strands that make up each strand of the rope itself are unlaid. Take a strand from one side and two from the other, making three. Then seize them together as shown in the illustration. They are now cut off close to the seizing.

FIG. 76A and B: The *Three-Strand Sailmaker's Short Splice* is begun by marrying the strands of each rope. Choose any two strands lying next to each other, in order to have one adjacent strand from each rope. Then instead of going over one and under one, as in the common Splice, tuck them around one another. As there is a right and wrong way of tucking these strands in the Sailmaker's Splices, the following method should be followed to eliminate the possibility of making any errors: when the strands have been married, select the proper strand with which the first tuck is to be made; look closely at the strand it is to be tucked under, and observe the direction the yarns of that strand follow; then tuck the strand in the same direction as the yarns, or in the direction of strand *a*. FIG. 76B illustrates the finished Splice, after three rounds of tucks have been made and the temporary whipping and strands cut off.

FIG. 77: The *Four-Strand Sailmaker's Short Splice*.

Plate 28—Three-Strand Long Splice

FIG. 78A: The *Three-Strand Long Splice* (over and under style), though weaker than the Short Splice and requiring more rope, does not appreciably increase the diameter of the rope, and can therefore be run over a sheave or pulley without jamming in the block. To begin, unlay the strands of both ropes four to five times farther than in a Short Splice. (In reality, the strands are unlaid much farther than the specimen in the illustration, since it was necessary to make the Splice shorter in order to get the work into the photograph.) Then marry the strands as shown here, and group them off into pairs. The next step is to take strand *a* and unlay it a short distance. Then take its mate, strand *b*, and lay it into the groove that was made when strand *a* was removed.

When strand *b* has been laid up to strand *a*, and they meet, tie an Overhand Knot with both strands to hold them temporarily. Next, go back to the center, and knot strands *c* and *d* together, as they will remain where they are. Take strand *b* and unlay it to the left the same distance that strand *a* was unlaid to the right. Then lay strand *e* into the groove left when strand *b* was taken out.

B: This shows how the Splice will look after the preceding steps have been made. (The Overhand Knots have not been tied in the strands in this illustration.) Strands *a* and *b* are together on the right; and *c* and *d* are together in the center. Strand *f* has been unlaid, and strand *e* is being laid into the groove until it reaches strand *b*.

FIG. 79: *Finishing Off the Strands of a Long Splice*. The next and final step in long splicing shows the disposal of the strands. This can be accomplished in several ways. The method illustrated shows the whole strand Overhand Knotted. (Note the manner in which the strands are knotted together, with the strands coming out on the right on top, and coming out on the left on the bottom. When tied in this fashion, the strands lie flat and do not form a lump.) After they have been knotted together, splice them over and under against the lay, as in the ordinary Short Splice. This is done to each pair of three strands.

FIG. 80A: *Split Strand Method*, First Method, illustrates another method perhaps not as strong as the one described in FIG. 79, but forming a neater joint. It is made by first splitting the strands in half, then tying an Overhand Knot with the halved strands,

as indicated. In this case, the splicing strands, *a* and *b*, are the ones with which the Overhand Knots were formed. Strands *c* and *d* are left as they are until the Splice is completed. Then after the Splice has been rolled under the foot, they are cut off, leaving about an inch of each strand, so that when the Splice is stretched under tension the tucks will not pull out. Either an Over and Under or Sailmaker's Splice may be made in finishing off the strands.

B: This shows the Splice before the strands have been pulled up taut. Strands *a* and *b* are the strands with which the Overhand Knots are tied, and *c* and *d* are the dead strands to be cut off.

FIG. 81A: *Split Strand Method*, Second Method, shows another method in which the strands are first split and knotted together as in FIG. 80. Then take one strand and tuck it under the next one to it. In other words, the strand that comes out on the right is tucked under the next strand to its right, and the strand on the left is tucked under the next strand to its left. This illustration shows working-strand *a* already tucked, and *b* tucked halfway. The dead strands *c* and *d* are left until the Splice is finished, and then cut off.

B: The strands will now emerge from the other side of the rope, to be spliced in as a Sailmaker's Splice.

C: This shows the Splice after two tucks have been made. In a Splice of this type, it is good practice to put in at least four tucks with each strand.

D: The Splice in this illustration is shown as it looks when completed.

Plate 29—Six-Strand Long Transmission and Cable-Laid Splices

FIG. 82: The *Four-Strand Long Splice*. All strands were disposed of in this Splice by employing the method illustrated in PLATE 28, FIG. 81.

FIG. 83: The *Six-Strand Long Splice* is begun by seizing the rope the proper distance from the end, or about twice as far as in three-strand rope. Unlay the strands

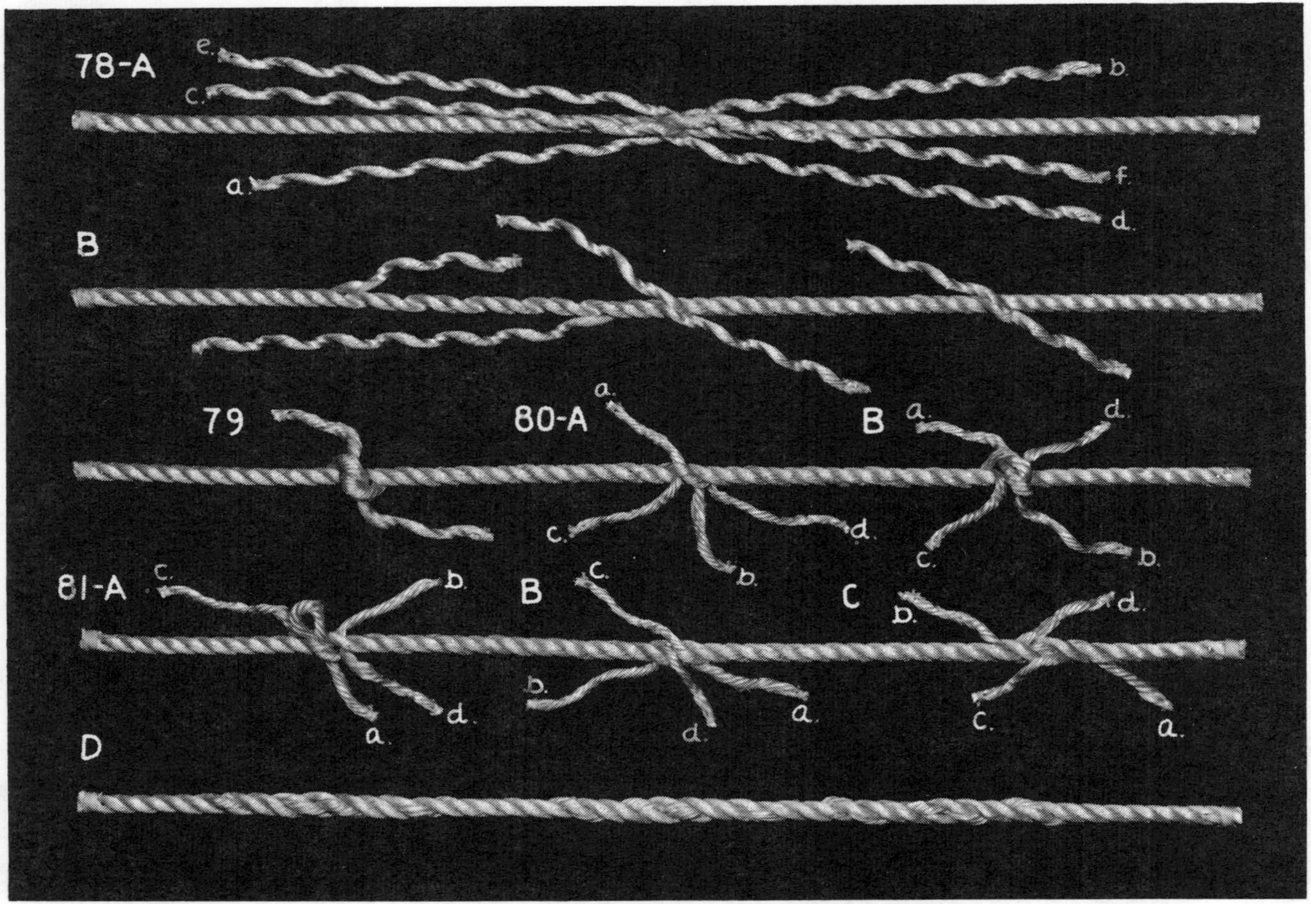

PLATE 28—THREE-STRAND LONG SPLICE

of both ropes to the seizing, cut out the hearts, and marry the strands. Select alternate strands on one rope, leaving three strands. Put a temporary seizing on these strands, to hold them secure to the standing part of the rope. Then cut off the seizing on the rope from which the three strands were selected, and taking the strands in pairs, proceed as in making a Three-Strand Long Splice. After this has been done, the seizing is cut off the other rope, and the same procedure is followed.

FIG. 84A: The *Transmission Splice* is similar to the Long Splice, but the manner of finishing off the strands is quite different. Whereas the common Long Splice will stand up when used on hand tackle, the strands would work loose if it were used as a Transmission Splice. (A Transmission Splice is tied in factories on ropes used to transmit power from motors to machinery, and is therefore subjected to high speeds.) This Splice, as a rule, is made in four-strand rope; but it can also be tied in three or six-strand rope. The first step is the same as in the common Long Splice. The difference lies in the method of tucking and finishing off the strands. After the strands have been laid up in their proper positions, take one strand—the one on the left-hand side—and unlay it three more full turns. Split the strand and name one half *a*, the other half a^1. Next, take the half-strand *a* and lay it up three full turns, until it reaches its original position. Take its partner, and unlay this strand to the right three full turns. Split the strand in half and lay it up to its original position, naming one half *b* and the other half b^1. In this illustration, strand *b* has been laid up to its original position, but strand *a* has not. (NOTE: as space was limited in the photograph, it was not possible to unlay the strands three full turns as required. But although they were unlaid only one turn, the principle is the same.)

B: When halves *a* and *b* meet, they are joined with an Overhand Knot in the usual manner. Tuck the half-strand *b* to the left, around half-strand *a* as in a Sailmaker's Splice, until *b* reaches a^1. The half-strand *a* is then tucked to the right around half-strand *b*, until b^1 is reached. This illustration shows *b* tucked until it has reached a^1. The half-strand *a* has not yet been tucked to reach b^1.

C: When the preceding step has been completed, take strand a^1 and open the strand in the center, close to the rope itself. Then insert strand *b* through this hole. After this has been done, it is tucked through the center of the rope, emerging from the opposite side. Strands b^1 and *a* show how this is handled.

D: The strands are now all drawn taut and cut off, leaving about an inch of each strand; for when the rope is stretched a small amount of slack will be taken up. After the rope has been run in for several days, the small remaining pieces of yarn from each strand may then be cut off.

FIG. 85: The *Three-Strand Transmission Splice*. All strands are disposed of in three-strand rope by using the same method as used in four-strand rope.

FIG. 86: The *Long Splice (Cable-Laid Rope)* or *Mariner's Splice* is tied by first unlaying the rope about two and one-half times the length that would be required for an ordinary Long Splice in a rope of the same size. Marry the strands and proceed as in the common Long Splice. When they have been spaced apart the proper distance, unlay the strands of one pair, marry them, and put in the common Long Splice. Then repeat the same procedure with the other two pair. Patience, perseverance, and a substantial amount of elbow-grease will be found necessary to turn in a Splice of this type neatly and securely.

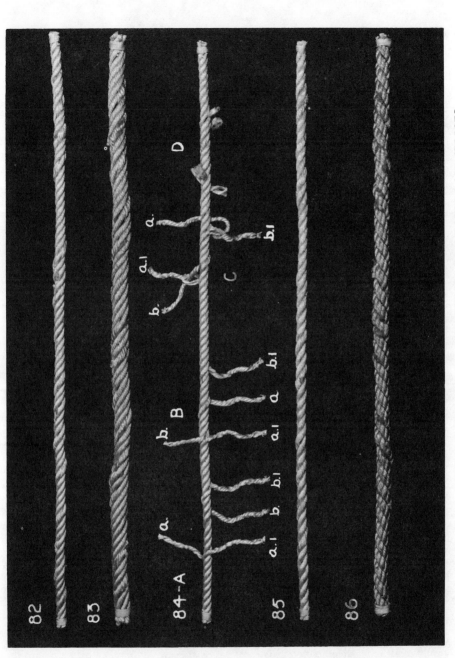

PLATE 29—SIX-STRAND LONG TRANSMISSION AND CABLE-LAID SPLICES

Plate 30—Rope Lengthening Splices

Fig. 87A: The *Splice for Lengthening a Rope with a Single Strand* is useful whenever it is desired to give a sail more spread by inserting a cloth. The head and foot rope can be lengthened in the following manner: cut a strand at the point marked *a;* unlay the cut strand to point *b,* and cut another strand; then unlay both cut strands to point *c,* and cut the last strand. By following this method, the wrong strand cannot possibly be cut.

B: Next, measure the distance from *x* to *z.* This will be the distance that the rope is to be lengthened. (Be certain to leave enough end on strand *2* for splicing.) Lay strand *2* up to the point *z.* Measure off the same distance on the other rope, and lay strand *b* up on strand *a.* Mate strand *a* with strand *2,* and lay strand *3* up until it mates with strand *b.* Select a strand from another rope of the same size, and measure it off to be at least eight inches longer than the dis-

tance from strand *c* to strand *1.* Take this strand and begin laying it in from strand *c* until it reaches strand *1.* The single strand is lettered *d.*

C: This shows how the Splice will appear after the steps just outlined have been completed. All that remains now is to splice the strands in the usual manner.

Fig. 88A: The *Splice for Lengthening a Four-Strand Rope* is accomplished in almost the same fashion as the Three-Strand Splice. Join strands *a* and *2,* and strands *b* and *3.* Strand *4* is then laid up on the left-hand rope until it mates with *c.* One long single strand is then taken from a four-strand rope of the same size, and laid in from strand *d* to strand *1.* This long strand is lettered *e.*

B: When the additional strand has been added, the Splice appears as in this illustration. Then splice the strands in the usual way.

Plate 31—Miscellaneous Splices

Fig. 89A and B: The *Grecian Splice* must have the rope unlaid twice the distance required for the common Short Splice. Then from each strand on both ropes take four yarns with which to form nettles for the cross-pointing. Lay up enough of each strand again so that a Short Splice with three rounds of tucks can be made. Marry the ropes, and make a Short Splice. After this has been done, bring each strand up between each group of nettles, and seize them temporarily to the standing part of the rope. Next, take each nettle, or yarn, and separate it from the next one in its group, until there are three sets of four yarns. Twist these down around the Splice in the same direction as the lay of the rope. When they meet the group of nettles from the other rope, lead them between the groups and seize them temporarily, just behind the second group. Unlay the second

group in the same manner as the first. Then in groups of four, tuck them over four yarns and under four of the first group, just as in splicing, and against the lay. When the end of the Splice is reached, scrape-taper all the strands and marl them down well. Fig. 89B shows the finished Splice.

Fig. 90A and B: The *Eye Splice Wormed and Collared* is a very smart method of turning in an Eye Splice, and may be used on a rope that does not come close to a block. Unlay the rope about twice as far as in the common Splice. When the first tuck has been made, take four yarns from each strand, and with the remaining yarns make a Sailmaker's Eye Splice, tapering by leaving out several yarns at intervals. Next, take two yarns from each pair of four nearest the eye, and twist them up into two yarn nettles. These are then wormed into

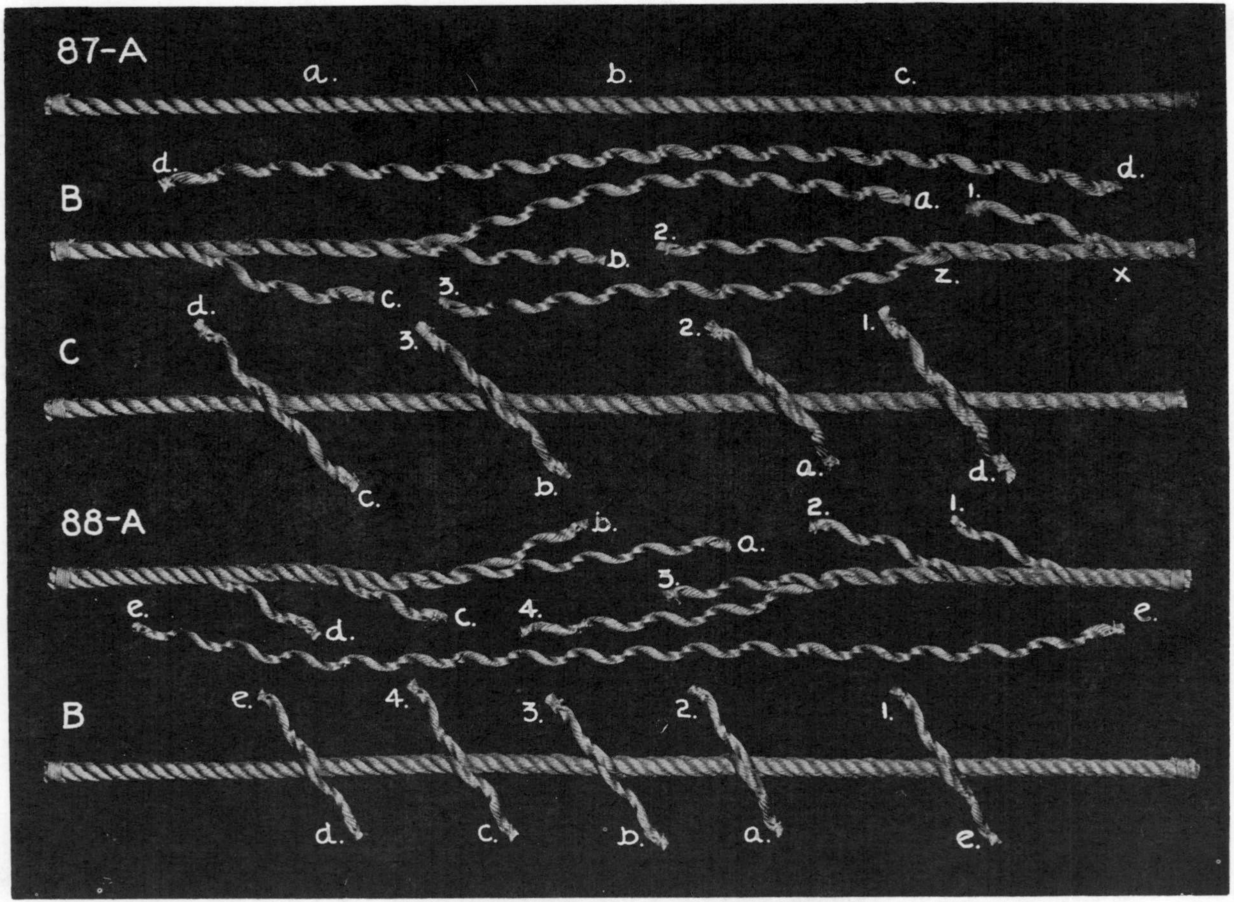

PLATE 30—ROPE LENGTHENING SPLICES

the lay until they reach the end of the Splice, where they are tucked under one strand. The remaining yarns, named group *a,* are twisted into three two-yarn nettles, and a Footrope Knot is formed. Three passes are then made, the knot is drawn taut, and the ends are cut off short. Select two yarns, *b,* from each of the remaining strands of the Splice (not the worming nettles), and cut off all the other yarns. Roll each group of two yarns into nettles and form the second Three-Pass Footrope Knot. Draw it up taut, and cut the nettles off short. Fig. 90B shows the knot complete.

Fig. 91: The *Running Eye Splice on the Bight* is begun by twisting the rope in the middle, against the lay. The strands will then untwist spontaneously and begin twisting up on themselves. When they have one complete twist in them, they are spread apart, and the end rove through the eye of each strand.

Fig. 92: The *Ropemaker's Eye* is a type of eye that is formed only by people who manufacture rope. Instead of beginning to twist a cable-laid rope from the end, a long piece of rope is taken (in cable-laid rope, this would form two of the three strands of the rope), and the bight of this is taken. The third strand is bent to the size of eye desired. There is now one single end with an eye and one bight, which forms three strands. These three are twisted to make the rope. After a suitable length has been made, strands *a, b,* and *c* are wormed into the lay, instead of being spliced, and the eye and worming are then marled and served with good rope.

Fig. 93: The *Admiral Elliott's Eye* is similar to the Ropemaker's Eye, but no loop is formed by the bight of two of the strands. Instead, the loop is formed by taking two strands from the end of the rope and long-splicing them together. The third strand is bent around, and an eye of the proper size is spliced in. Then instead of cutting off the strands remaining from the Eye Splice, they are wormed into the lay of the rope, and the whole eye and worming are marled and served with good rope.

Plate 32—Cringles and Miscellaneous Eye Splices

Fig. 94A: The *Cringle,* like the Grommet, is made with a single strand of rope. Instead of being made by itself, it is worked into another rope, such as the leech rope of a sail. Take one strand of a three-strand rope, and stick one end through the rope and under two strands. Be sure that end *a* is toward you and end *b* leading away from you. In other words, stick the single strand against the lay of the rope. Also make certain that there are an odd number of kinks in the part of the strand that is to form the Cringle. The one shown in this illustration has three: *c, d* and *e.*

B: Selecting end *a,* proceed to lay it up on its own bight until it reaches the opposite strand.

C: Now take end *b,* and lay this strand up to form a three-strand rope.

D: The ends are now disposed of as in a common Splice. This illustration shows the completed Cringle.

Fig. 95: The *Cringle with a Thimble* is begun in the same manner as the Cringle described in Fig. 94, but using the strand from a four-strand rope. Lay the strands up as in the Three-Strand Cringle. When they are laid up until they meet the rope, they are rove through the rope again in the same direction they were first put through. They are then laid up again until they meet at

NOTE: Some authorities prefer to form the Cringle first and then hammer the thimble into position. This is a very good method, although considerable experience is necessary in order to avoid making the Cringle too small or too large.

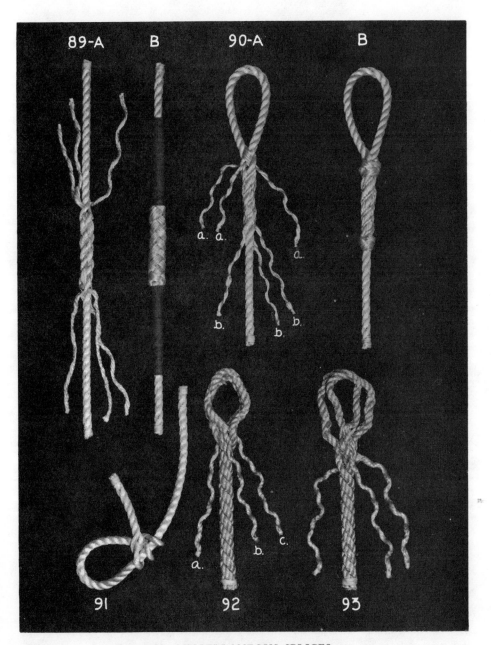

PLATE 31—MISCELLANEOUS SPLICES

the top of the thimble. The strands are next hove taut with the aid of a marlin-spike, and the ends cut off short.

Fig. 96a: The *Flemish Eye Splice,* which is a little different from the common Eye Splice, can be used when there is a tendency to spread the eye apart. First, unlay the rope a sufficient length for turning in a common Eye Splice. Select one of the three strands and unlay it until the remaining two strands form an eye of the desired size. Bend these two strands down until they meet the single strand *a* again. Instead of laying it up on the standing part, lay it in the other direction, beginning a short distance from the end of the opening strands.

b: When strand *a* has been laid up again, proceed to dispose of the ends as in the common Splice.

Fig. 97: The *Flemish Eye Splice* (four strands) starts as usual with one strand. When this has been completed, select an-other strand, but not one adjacent to the first one used. Then splice in the strands as before.

Fig. 98a and b: The *Eye Splice on the Bight* is begun by untwisting the rope where it is desired to make the eye. Continue doing this, and the strands will untwist on themselves as shown. When these ends are long enough, form an eye and proceed to splice them into the rope, following the method used for a Three-Strand Splice. Fig. 98b shows the finished Splice.

Fig. 99: The *Eye Splice on the Bight* (four strands).

Fig. 100: The *Combined Horseshoe and Eye Splice* has the strands unlaid about four times more than in an ordinary Splice. Then splice an eye in the end of the rope. After three tucks have been made, lay the strands up again to form the horseshoe. Splice the remaining strands into the standing part of the rope.

Plate 33—Miscellaneous Splices

Fig. 101: The *Shoemaker's Splice* can be used in either old rope or rope with a loose lay. Marry the rope in the usual manner, and make one round of tucks against the lay of the rope. The next round of tucks is made with the lay, or sailmaker fashion, and the last round is again made against the lay. That is, one against, one sailmaker style, and the last against. Roll the Splice, and cut the strands off short.

Fig. 102: The *Four-Strand Shoemaker's Splice.*

Fig. 103: The *Drawing Splice* is a good way to short-splice a hawser or large rope, and enables one easily to pass the spliced portion of the line through a hawse-hole or chock. By using this method, the Splice may also be taken apart again without damage. Put a stop on each hawser, about two fathoms from the end, depending on the size of the rope. Each rope is unlaid to the stop, and another stop placed on each strand about four feet from the end. The strands are then scrape-tapered and laid up again. Each strand of both ropes is next worked into a long taper. Marry the hawsers, and short-splice them in the usual manner. When the end of the Splice has been reached, clap a stout seizing on each side. The tapered ends are now wormed around the cable. When half of the strand has been wormed, clap on another seizing. Then continue worming until the ends of the strands are reached. At this point, clap on another strong seizing. The finished Splice is shown in the illustration.

Fig. 104a and b: The *Splice for Adding a New Strand to a Rope* is often necessary when a strand has become frayed. Pick up the frayed strand and cut it in two. (Do not cut the entire rope in two—only one strand.) Unlay both cut ends a good distance on each side of the frayed portion.

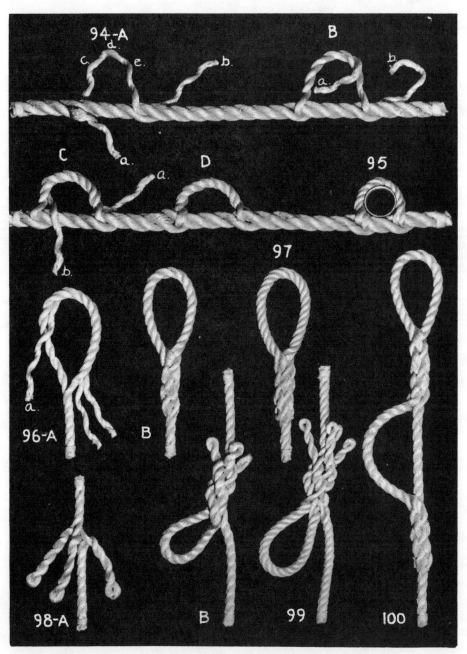

PLATE 32—CRINGLES AND MISCELLANEOUS EYE SPLICES

Then take a single strand from another rope of equal size, and lay it into the groove from which the cut strand was unlaid, leaving enough end with which to splice. Then dispose of the ends as in the common Long Splice. Fig. 104B shows the Splice completed.

Fig. 105: *Splicing Three-Strand and Four-Strand Rope Together* is as follows: unlay the strands of both ropes a sufficient distance, and marry them: unlay one strand of the three-strand rope, and fill the space with a strand of the four-strand. Next, unlay a strand from the four-strand, and fill the space with a strand from the three. There now remain two strands from the four, and one strand from the three. Divide the single strand from the three, by taking out one third. Knot this third to one of the remaining pair of the four. Then unlay the remaining strand of the four, and fill the space with the two thirds left from the three-strand rope. Finish off the strands as in the common Long Splice. This shows how the Splice appears before the ends are tucked. Another method is to work three strands as usual, and splice in the remaining fourth strand where it lies. The first method is the better.

Fig. 106: The *Splice for Shortening a Rope* is begun by cutting the rope in the same fashion as though you were preparing to lengthen it. Then marry the ends which were cut, and splice the strands in the usual way.

Fig. 107: The *Spectacle Splice* has the strands laid up in the same fashion as the Grommet (Plate 43, Fig. 161). It should be done in such a way that when finished,

two strands of one will run under two strands of the other. This Splice is made of four strands. Finish off the ends as in a Long Splice.

Fig. 108: The *Twin Eye Splice* is made in the same manner as the preceding knot, but with three strands.

Fig. 109: The *Eye Splice in Eight-Strand Braided Hemp* or *Signal Halyard Splice* is made in the following way: unbraid a sufficient amount of the end, and separate the strands into two groups of four strands each; bend the end around until an eye of the desired size has been formed. Then take a pricker and tuck the four strands on the right-hand side, so that each one follows a strand in the standing part. The same is done with the strands on the left, in the opposite direction. Follow these same strands in the standing part until the desired amount of tucks have been made. Four tucks should be made if this Splice is to carry considerable weight.

Fig. 110: The *Grecian Eye Splice* requires about three times the amount of rope necessary for a common Splice. After making the first tuck, take eight yarns from each strand and seize them to the eye, to keep them out of the way temporarily. Continue with the common Splice, tapering the strands. Next, remove the temporary seizing and divide the strands so that there will be six sets of four strands each. Begin to cross-point, alternately taking three sets to the right and three to the left. When the entire Splice has been covered, put a stout seizing over all six sets of yarns, in this way securing them to the standing part of the rope.

Plate 34—Miscellaneous Splices

Fig. 111A and B: The *Horseshoe Splice* is begun by unlaying a short piece of rope as shown in Fig. 111A, and splicing it into the standing parts of another piece of rope which has been bent in horseshoe fashion.

The completed Splice will then appear as in Fig. 111B. This Splice was formerly used to separate the legs of a pair of shrouds.

Fig. 112: The *Sailmaker's Short Splice in Three-Strand Grommet* first has the

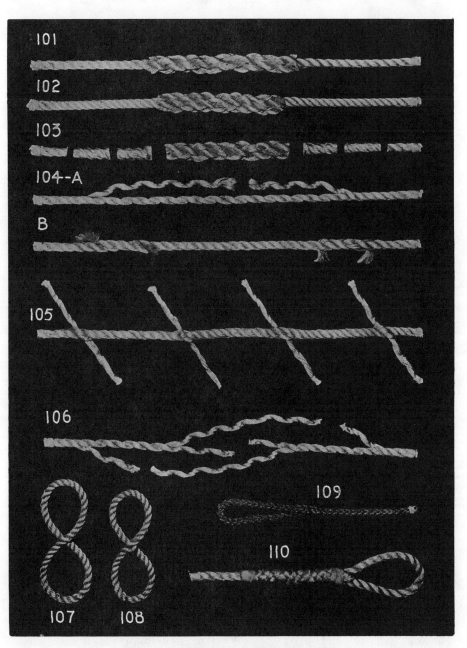

PLATE 33—MISCELLANEOUS SPLICES

strands married together. Then an Overhand Knot is tied with each set of strands, which will bring the strands out in the form of a sailmaker's tuck. Continue with sailmaker's tucks and taper off. This Splice is seldom used.

FIG. 113: The *Sailmaker's Short Splice in Four-Strand Grommet.*

FIG. 114: The *Combination Three-Strand Splice*, First Method, is begun by first marrying the ropes. Then with each set of strands make a sailmaker's tuck in the form of an Overhand Knot. Make the next set of tucks as in a regular Splice (against the lay). Finally, taper down, and cut the strands off short.

FIG. 115: The *Combination Four-Strand Splice*, First Method.

Plate 35—Braided Eye, Horseshoe, and Cargo Strap Splices

FIG. 116: The *Three-Strand Braided Eye Splice* is made by taking the first set of tucks, as in a Three-Strand Eye Splice, and then separating the strands into braids of three strands each.

FIG. 117: The *Stirrup Splice with Eyes* is made by splicing one line into the standing part of another line, with spliced eyes in both ends. By splicing hooks in the eyes, they can be used as a Barrel Sling or a Double Sling.

FIG. 118: The *Four-Strand Braided Eye Splice* is formed by first making a set of tucks, as in a Four-Strand Eye Splice, then separating the strands and making four-strand round braids out of each set.

FIG. 119: The *Horseshoe and Crossed Loop Splice* shows a Loop Splice crossed over and seized into the standing part of a Horseshoe Splice.

FIG. 120: The *Cargo Strap* can be made in various sizes, according to requirements. For an extra load, two straps can be used by splicing the loop of one over the loop of the other. Then pull the first strap through the loop. Do not taper these Splices, and use several additional tucks on each end. To shorten a strap, tie an Overhand Knot with the two loops under the standing parts.

Plate 36—Cut Splice and Eye Splice with French Seizing

FIG. 121: The *Eye and Loop Splice with Horseshoe Splice in the Standing Part.*

FIGS. 122A and B: The *Cut Splice* is made by measuring off the required distance, then splicing each end into the standing part of the other rope. FIG. 122A pictures the rope in position for starting; and FIG. 122B shows the completed splice. The Splice is used to form an eye or collar in the bight of a rope.

FIG. 123: The *Three-Strand Eye Splice with French Seizing* consists of an eye spliced in, with the ends frayed out and seized to the standing part of the rope. This is a neat way of finishing off an Eye Splice in a hawser. It will also prevent the spliced portion of the rope from jamming in a hawse hole.

FIG. 124: The *Loop Splice* is a loop spliced into the standing part of a rope.

Plate 37—Miscellaneous Splices

FIG. 125: The *Eye Sling* or *Strap Splice* is made as follows: unlay the end of a line the required distance, and tuck the ends to the size of eye desired; then make about three full tucks, and lay up the strands. First, lay up two of the strands, seize the ends, and lay up the third one. Finish off by making a Short Splice with both ends.

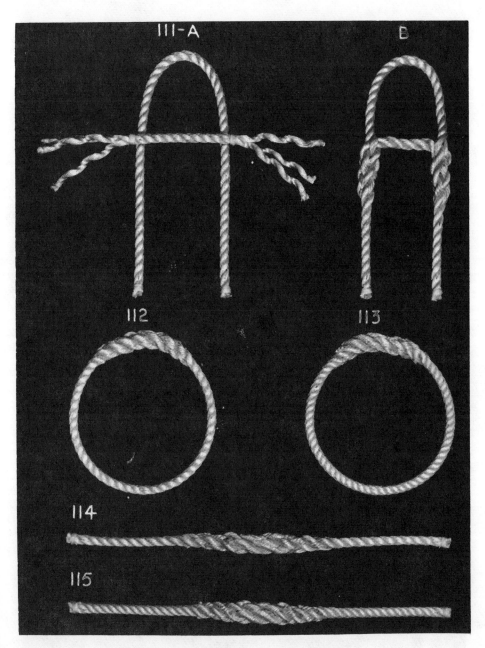

Plate 34—MISCELLANEOUS SPLICES

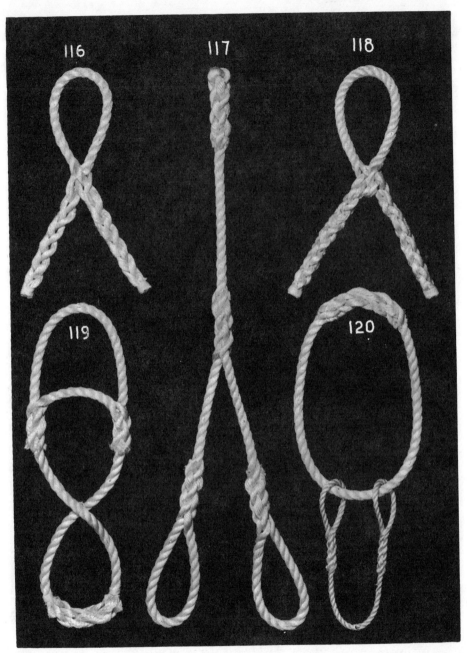

PLATE 35—BRAIDED EYE, HORSESHOE, AND CARGO STRAP SPLICES

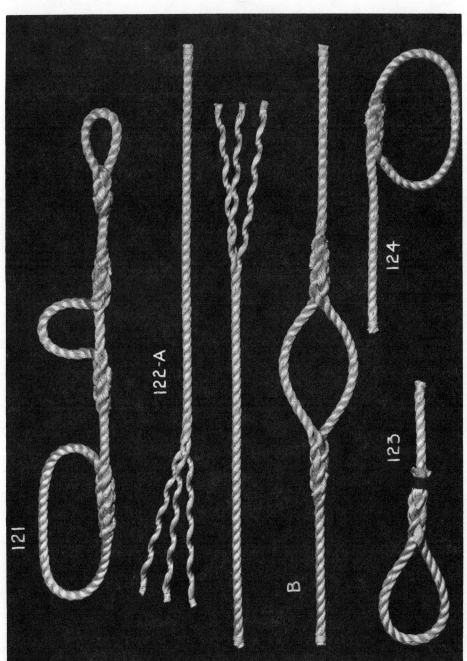

PLATE 36—CUT SPLICE AND EYE SPLICE WITH FRENCH SEIZING

Fig. 126: The *Loop and Eye Splice Halter* represents a halter formed with Loop and Eye Splices.

Fig. 127: The *Eye and Loop Splice* is an eye and loop spliced into the standing part of a rope.

Fig. 128: The *Short Splice Served* consists of an ordinary Short Splice tapered and served.

Fig. 129: The *Six-Strand Sailmaker's Short Splice*.

Plate 38—Rope Halter and Miscellaneous Splices

Fig. 130: The *Three-Strand Single Stopper Shroud* is made up of two Single Crown and Wall Knots tied back-to-back in the usual manner.

Fig. 131: The *Emergency Rope Halter* is made as follows: at the required distance from the end of the rope, form a Man-Harness Knot; then place the long end of the rope around the knee, with the short end on top, and join the short end to the long end with an Englishman's Tie (two Overhand Knots arranged in the manner shown). Pull this coupling up tight, and pass the long end through the eye of the Man-Harness Knot.

Fig. 132: The *Four-Strand Single Stopper Shroud*.

Fig. 133: The *Three-Strand Single Manrope Shroud* consists of two Single Wall and Crown Knots tied back-to-back in the usual fashion.

Fig. 134: The *Combination Eye and Back Splice in Slip Loop* shows a small Flemish Eye in the form of a Slip Loop. A Back Splice is tucked in with a regular Eye Splice, and each set of strands laid up and seized.

Fig. 135: The *Four-Strand Single Manrope Shroud*.

Plate 39—Sailmaker's Eye and Combination Splices

Figs. 136A and B: The *Three-Strand Sailmaker's Eye Splice*, Second Method, has the eye bent in from left to right with the lay, instead of from right to left against the lay as in the First Method (Plate 20, Fig. 11). Make the first tuck with the strand nearest the standing part, tucking it under one strand with the lay. Follow up with the other strands in rotation, making each tuck under one, and then continuing with the regular sailmaker's tucks. The eye will appear as in Fig. 136A when the first set of tucks has been completed. Fig. 136B shows the Splice when finished. This type of Sailmaker's Eye Splice lays in better and looks neater than the regular method, which has the first tucks made against the lay.

Figs. 137A and B: The *Four-Strand Sailmaker's Eye Splice*, Second Method, follows the three-strand method, except that the first tuck is under two instead of one. Fig.

137A shows the first set of tucks completed. Fig. 137B shows the Splice as it looks when finished.

Figs. 138A and B: The *Six-Strand Sailmaker's Eye Splice*, Second Method, is begun by unlaying the strands a sufficient length. Then seize them and remove the heart. Bend the eye in toward the standing part from the left, or with the lay, as in the three- and four-strand methods. Tuck the strand nearest the standing part under three with the lay; tuck the next or following strand under two; and continue by tucking each of the remaining strands under one. In other words, the tucks are the same as in the Liverpool Eye Splice in wire. The remaining tucks are over and under with the lay, as in the regular Sailmaker's Eye Splice. Fig. 138A has the first tuck completed, and Fig. 138B shows the Splice when finished.

Fig. 139: The *Three-Strand Lock Seiz-*

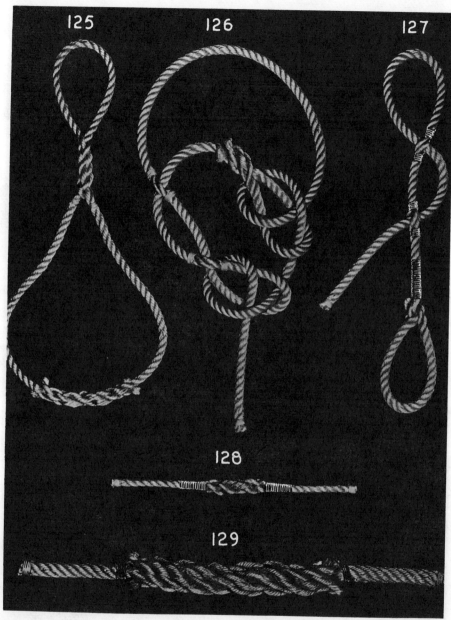

Plate 37—MISCELLANEOUS SPLICES

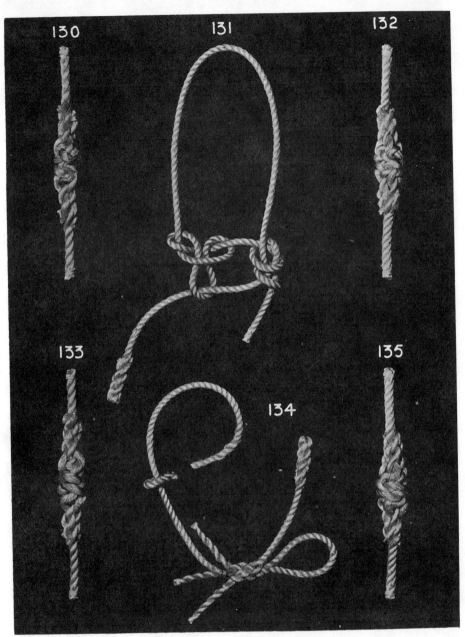

PLATE 38—ROPE HALTER AND MISCELLANEOUS SPLICES

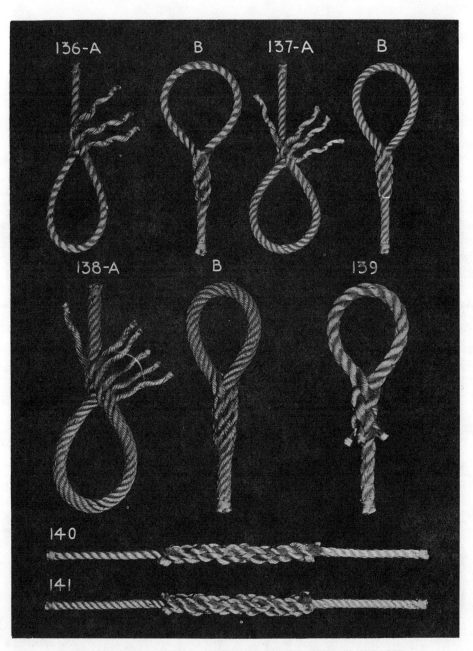

PLATE 39—SAILMAKER'S EYE AND COMBINATION SPLICES

ing in Hawser is begun by splicing the eye in with the required number of tucks. Then finish off by splitting the strands and seizing them together in pairs. This is done to prevent the strands from working out, as will happen when working with large rope.

Fig. 140: The *Three-Strand Combination Splice,* Second Method, starts with three tucks made against the lay after joining the ropes together. Then the next three tucks are made with the lay to complete the Splice.

Fig. 141: The *Four-Strand Combination Splice,* Second Method.

Plate 40—Halters and Bull Earing Splices

Fig. 142: The *Bull Earing Splice* can be easily and simply tied by following a method common in the days of the sailing-ships. An old and well-worn piece of manila was spliced into the standing part, forming a long bight in order to hitch around the yard outside the cleat. Then it was rove through the reef cringle and back to the yard, where it remained, instead of being left in the cringle.

Figs. 143A and B: The *Loose Guard Loop Halter* is tied as follows: form the eye by making the first tuck with the long end of the rope, under two strands and toward the left as indicated; then tuck the short end of the rope under two strands of the long end, also toward the left, skipping about two strands before making the tuck. Bring the long end around to form the nose-piece, and tuck it through two strands of the short end, which is brought over from the eye on the right to form the head-piece. Tuck the short end through two strands of the long end, toward the right, passing it through the eye before making it secure by tying an Overhand Knot around the long end and through its own part. Long end and short end are represented as *a* and *b*.

Fig. 144: The *Bull Earing Splice with Twin Bights* is a type that was sometimes made to give more parts in the first turn, by splicing an additional length into the first bight. The bights, however, had a tendency to twist up in wet weather, so the simpler form was usually preferred.

Fig. 145: The *Served Grommet* is a regular Grommet, served with Marline.

Fig. 146: The *Adjustable Halter* is begun by measuring off the proper length of rope according to the table below. Form an eye with the part of rope used for the long bight or nose-piece. Tuck it under one strand of the other part of rope representing the head-piece. Tuck the head-piece under two strands of the nose-piece, toward the right, to complete the eye. Bring the nose-piece around to form the proper length of bight, and splice it into the head-piece. Adjust the halter to the animal, and secure the lead to the loop if it is desired.

TABLE FOR ADJUSTABLE HALTER

	Diameter of rope	Total length of rope	Length left for lead
Large cattle	½ inch	12 feet	6 feet
Medium cattle ..	⅜ inch	11½ feet	6 feet
Small cattle	⅜ inch	11 feet	6 feet
Calves and sheep.	¼ inch	7½ feet	4 feet

Fig. 147: The *French Served Grommet* is a regular Grommet, served with French or Grapevine Hitching.

Plate 41—Miscellaneous Eye and Shroud Splices

Fig. 148: The *Four-Strand Eye Splice with French Seizing* is made by splicing the eye in, and then fraying the ends, which are finally laid around the standing part and seized.

Fig. 149: The *Regular Short Splice in Grommet.*

Fig. 150: The *Four-Strand Eye Splice with Lock Seizing* is tied by first splicing in the ends to form the eye. After the re-

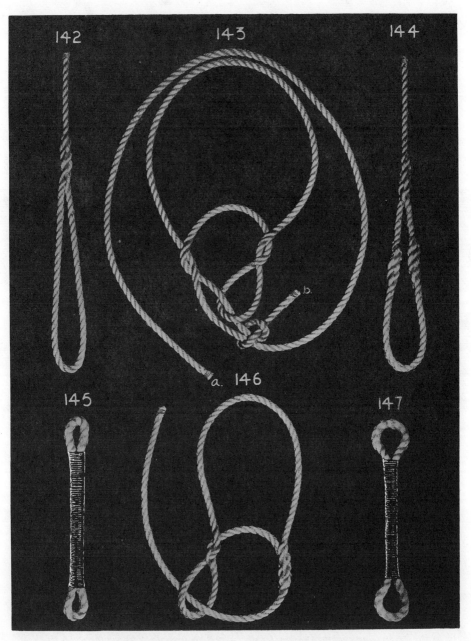

PLATE 40—HALTERS AND BULL EARING SPLICES

quired number of tucks have been made, each strand is split in half and seized to the opposite half of the next strand. This brings them all together in sets of two halves each.

FIG. 151: The *Three-Strand Inverted Wall Shroud* is composed of two Inverted Wall Knots, tied back-to-back in the usual manner, and spliced into the standing part of the rope.

FIG. 152: The *Four-Strand Inverted Wall Shroud.*

FIG. 153: The *Eye Splice and Thimble Served* has an Eye Splice made around a thimble, then tapered and served.

FIG. 154: The *Three-Strand Double Walled Crown Shroud* is made up of two Walled Crown Knots, which are doubled and tied back-to-back in the usual way, and then spliced into the standing part of the rope.

FIG. 155: The *Four-Strand Double Walled Crown Shroud.*

FIG. 156: The *Double Sheet Bend in Spliced Eye.*

Plate 42—Adjustable Halter and Chain Splices

FIG. 157: The *Non-Adjustable Halter* is formed by tucking the head-piece under one strand of the nose-piece, toward the left. Then skip two strands, and bring it back under one strand of the nose-piece, toward the right, to form the eye. After measuring the required length of bight for the nose-piece, splice the head-piece into the nose-piece at the proper distance. Splice one strand up toward the top of the bight and one strand down toward the bottom, with the other strand being spliced back into the head-piece.

FIG. 158: The *Standard Guard Loop Halter* is similar to the Loose Guard type, except that the short end is seized to the long end instead of being tied.

FIGS. 159A and B: The *Chain Splice* is tied as follows: unlay the rope a considerable distance, and reeve strands *b* and *c* through the end link of the chain; then unlay strand *a* quite some distance down the rope; tuck strand *b* under strand *c*, and lay up strand *c* in the groove vacated by strand *a*. Join them together with an Overhand Knot, disposing of the strands by tucking against the lay, as in a regular Long Splice, with strand *b* also tucked over and under against the lay to finish. FIG. 159A illustrates the open Splice at the beginning; and FIG. 159B shows how it looks when complete.

FIG. 160: The *Sliding Chain Splice in Loop.*

Plate 43—Grommet, Eye and Back Splices

FIG. 161A: The *Three-Strand Grommet* represents a type of knot that can be utilized for making many different articles, such as strops, chest handles, or quoits. It is made from a single strand of rope, as follows: first, take a piece of rope of the desired thickness; determine the circumference of the Grommet you wish to make; then measure off three and one-half times this circumference. (In other words, if the circumference is one foot, three and one-half feet is the proper measurement.) Next, cut the rope and unlay one strand. Tie an Overhand Knot in this strand, and begin laying up one strand as shown in the illustration.

B: When the ends meet, there will be a Grommet of two strands. Continue the strand around again, making three strands as illustrated here. (The strands have purposely been left short, in order to fit them into the photograph.)

C: When the strands meet again, halve them and tie an Overhand Knot. Then proceed to dispose of the ends as in the common Long Splice.

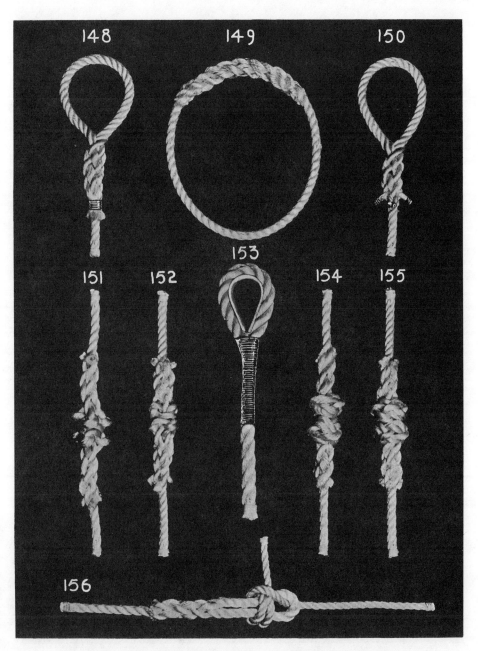

PLATE 41—MISCELLANEOUS EYE AND SHROUD SPLICES

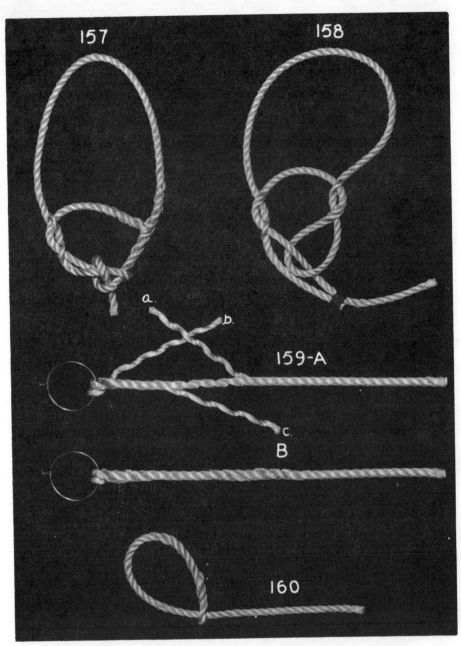

PLATE 42—ADJUSTABLE HALTER AND CHAIN SPLICES

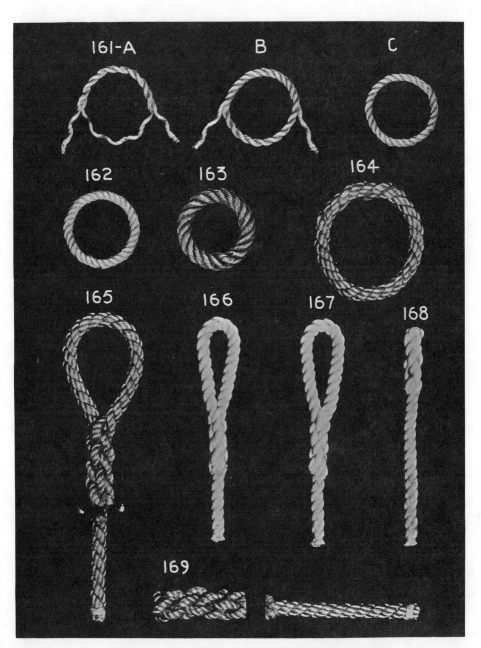

PLATE 43—GROMMET, EYE AND BACK SPLICES

Fig. 162: The *Four-Strand Grommet* is formed in precisely the same manner as the Three-Strand Grommet. But the strand used must be *four and one-half* times the circumference of the desired Grommet. The strand must of course be unlaid from a four-strand rope.

Fig. 163: The *Six-Strand Grommet* is tied by first making a heart slightly larger than the inside of the finished Grommet itself. This heart is also a Grommet, but of three strands. Next, take a strand of six-strand wheel rope, *eight* times the circumference of the completed Grommet, and proceed as before.

Fig. 164: The *Cable-Laid* or *Mariner's Grommet* resembles the common Three-Strand Grommet. When the strands meet for the last time, the three strands which comprise one strand of the cable-laid rope are unlaid in each strand, and married. They are then long-spliced together, and the strands disposed of in the usual manner.

Fig. 165: The *Cable-Laid* or *Mariner's Eye Splice* is begun by unlaying the three strands and proceeding to splice in the ordinary way. Four tucks are usually made. When finished tucking, split the strands and take one-half of one strand and one-half of another and seize them together as indicated.

Fig. 166: The *Sailmaker's Eye Splice in Cotton Line* is made in the same fashion as the Sailmaker's Eye shown in PLATE 20, Fig. 11. The strands in cotton line are a little more difficult to keep in place, and so should be tucked very carefully.

Fig. 167: The *Eye Splice in Cotton Line* follows the same method as the common Eye Splice.

Fig. 168: The *Back Splice in Cotton Line* is tied in the same way as the common Back Splice.

Fig. 169: The *Cable-Laid* or *Mariner's Back Splice* is formed in the same manner as the common Back Splice. When three tucks have been made, the strands are unlaid and seized to the standing part as shown. They also may be finished off as in the Eye Splice (Fig. 165).

NOTE: The strands in cable-laid rope should always be finished off in the manner shown; otherwise the strands may work free when towing with the Splice under water.

Fiber Rope Characteristics

In the table on the following page and on PLATE 44, are given the sizes and many of the characteristics of some of the different kinds of fiber rope most commonly used. Many sizes of Manila rope, other than those described here, are manufactured, usually for special applications, or for some specific purpose requiring that the rope have certain definite properties.

Fiber Rope Sizes and Characteristics

SIZES (Inches)		GROSS WEIGHTS (Lbs. and Decimals)		LENGTHS (Ft. and Decimals)		STRENGTHS (Pounds)	
Diameter	Circumference	Full Coils	Per 100 Ft.	Full Coils	Ft. in 1 Lb.	Breaking Strengths	Working Strains
3/16	5/8	25	1.43	1,750	70	450	90
1/4	3/4	30	1.72	1,750	58	550	110
5/16	1	45	2.63	1,700	38	950	190
3/8	1 1/8	60	3.71	1,625	27	1,300	260
7/16	1 1/4	75	5.26	1,425	19	1,750	350
1/2	1 1/2	90	7.5	1,200	13.3	2,650	530
9/16	1 3/4	125	10.4	1,200	9.6	3,450	690
5/8	2	160	13.3	1,200	7.5	4,400	880
3/4	2 1/4	200	16.7	1,200	6.0	5,400	1,080
13/16	2 1/2	234	19.5	1,200	5.13	6,500	1,300
7/8	2 3/4	270	22.5	1,200	4.45	7,700	1,540
1	3	324	27.0	1,200	3.71	9,000	1,800
1 1/16	3 1/4	375	31.3	1,200	3.20	10,500	2,100
1 1/8	3 1/2	432	36.0	1,200	2.78	12,000	2,400
1 1/4	3 3/4	502	41.8	1,200	2.40	13,500	2,700
1 5/16	4	576	48.0	1,200	2.09	15,000	3,000
1 1/2	4 1/2	720	60.0	1,200	1.67	18,500	3,700
1 5/8	5	893	74.4	1,200	1.34	22,500	4,500
1 3/4	5 1/2	1,073	89.5	1,200	1.12	26,500	5,300
2	6	1,290	108.0	1,200	.930	31,000	6,200
2 1/8	6 1/2	1,503	125.0	1,200	.800	36,000	7,200
2 1/4	7	1,752	146.0	1,200	.685	41,000	8,200
2 1/2	7 1/2	2,004	167.0	1,200	.600	46,500	9,300
2 5/8	8	2,290	191.0	1,200	.524	52,000	10,400
2 7/8	8 1/2	2,580	215.0	1,200	.465	58,000	11,600
3	9	2,900	242.0	1,200	.414	64,000	12,800
3 1/8	9 1/2	3,225	269.0	1,200	.372	71,000	14,200
3 1/4	10	3,590	299.0	1,200	.335	77,000	15,400
3 1/2	11	4,400	367.0	1,200	.273	91,000	18,200
3 3/4	12	5,225	436.0	1,200	.230	105,000	21,000

This table gives the sizes and characteristics of medium lay, three-strand Manila rope. The working strains are figured at about 20% of breaking strengths for efficiency in everyday service, but somewhat higher loads are not unsafe for temporary use.

Four-strand rope has approximately the same tensile strength as three-strand and runs 5 to 7% heavier.

Plate 44—The Relative Sizes of Various Ropes

FIG. 1. 3/16-Inch Size Three-Strand Manila Rope.

FIG. 2. 1/4-Inch Size Three-Strand Manila Rope.

FIG. 3. 5/16-Inch Size Three-Strand Manila Rope.

FIG. 4. 3/8-Inch Size Three-Strand Manila Rope.

FIG. 5. 7/16-Inch Size Three-Strand Manila Rope.

FIG. 6. 1/2-Inch Size Three-Strand Manila Rope.

FIG. 7. 1/2-Inch Size Four-Strand Manila Rope.

FIG. 8. 9/16-Inch Size Three-Strand Manila Rope.

FIG. 9. 5/8-Inch Size Three-Strand Manila Rope.

FIG. 10. 3/4-Inch Size Three-Strand Manila Rope.

FIG. 11. 13/16-Inch Size Three-Strand Manila Rope.

FIG. 12. 1-Inch Size Three-Strand Manila Rope.

FIG. 13. The Spun Yarn shown here is drab brown in color, and does not photograph very clearly.

FIG. 14. Eight-Strand Braided Hemp.

FIG. 15. The Old-Fashioned Six-Strand Wheel Rope is seldom seen nowadays in any country outside of France, where it is called "Septin" and still used extensively. The end is unlaid to show the heart.

FIG. 16. The Cable-Laid Rope is composed of three right-handed hawser-laid ropes laid up together left-handed, giving it nine strands in all. The end is unlaid to illustrate its method of construction.

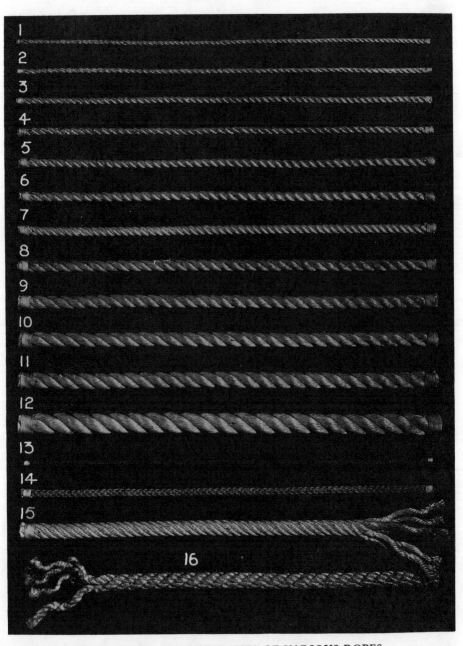

Plate 44—THE RELATIVE SIZES OF VARIOUS ROPES

Plate 45—Blocks and Tackles

As in any other mechanical hookup, there is a certain amount of friction loss in all tackles. It is difficult to determine the exact amount of the friction loss since it is based on many elements, such as: mechanical perfection of the blocks; relation of the rope to the sheave; and conditions under which the lift is being made, in regard to the lead of the movable block and hauling part.

Engineers have determined that, under conditions as nearly perfect as possible, the following ratios of power (P) to weight (W) will generally prevail:

Type		Friction not Considered	Friction Considered
No. 1:	*Single Whip*	$P = W$	$\dfrac{P}{W} = \dfrac{11}{10}$
No. 2:	*Single Whip* with block at weight	$\dfrac{P}{W} = \dfrac{10}{20}$	$\dfrac{P}{W} = \dfrac{12}{20}$
No. 3:	*Gun Tackle Purchase*	$\dfrac{P}{W} = \dfrac{10}{20}$	$\dfrac{P}{W} = \dfrac{12}{20}$
No. 4:	The same inverted	$\dfrac{P}{W} = \dfrac{10}{30}$	$\dfrac{P}{W} = \dfrac{13}{30}$
No. 5:	*Luff Tackle*	$\dfrac{P}{W} = \dfrac{10}{30}$	$\dfrac{P}{W} = \dfrac{13}{30}$
No. 6:	The same inverted	$\dfrac{P}{W} = \dfrac{10}{40}$	$\dfrac{P}{W} = \dfrac{14}{40}$
No. 7:	*Double Purchase*	$\dfrac{P}{W} = \dfrac{10}{40}$	$\dfrac{P}{W} = \dfrac{14}{40}$
No. 8:	The same inverted	$\dfrac{P}{W} = \dfrac{10}{50}$	$\dfrac{P}{W} = \dfrac{15}{50}$
No. 9:	*Spanish Burton*	$\dfrac{P}{W} = \dfrac{10}{30}$	$\dfrac{P}{W} = \dfrac{13}{30}$
No. 10:	*Double Spanish Burton*	$\dfrac{P}{W} = \dfrac{10}{50}$	$\dfrac{P}{W} = \dfrac{15}{50}$
No. 11:	*Bell Purchase*	$\dfrac{P}{W} = \dfrac{10}{70}$	$\dfrac{P}{W} = \dfrac{17}{70}$
No. 12:	*Luff upon Luff*	$\dfrac{P}{W} = \dfrac{10}{160}$	$\dfrac{P}{W} = \dfrac{26}{160}$

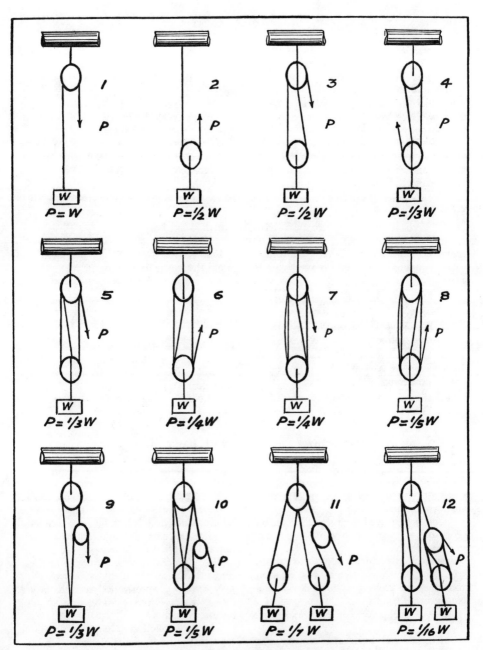

PLATE 45—BLOCKS AND TACKLES

To determine the power required to raise a given weight with a tackle, the general rule is: divide the weight to be raised by the number of parts of rope at the movable block or blocks, the quotient being the power required to produce a balance—friction not considered.

To find the amount of purchase required to raise a given weight with a given power: divide the weight by the power; the quotient will be the number of parts of rope to be attached to the lower block.

To find the weight a given tackle will raise: multiply the weight a single rope will bear by the number of parts at the moving block.

To find the amount of purchase gained when one tackle is put upon another: multiply the two powers together.

Remember that a considerable amount of friction loss must be dealt with and always make plenty of allowance for it when using these rules. Also remember: *power can only be gained at the expense of time.*

Plate 46—Blocks and Tackles

FIG. 1: *A Runner.* An additional power is gained when this is used with a purchase.

FIG. 2: *A Single Whip.* No power is gained.

FIG. 3: *A Double Whip.* Two times the power is gained.

FIG. 4: *A Gun Tackle.* Two or three times the power is gained, depending upon which is the movable block.

FIG. 5: *A Watch or Single Luff Tackle.* Three times the power is gained.

FIG. 6: *A Double Luff Tackle.* Five times the power is gained.

FIG. 7: *A Two-Fold Purchase.* Four times the power is gained.

FIG. 8: *A Single Spanish Burton.* Three times the power is gained.

FIG. 9: *A Double Spanish Burton.* Five times the power is gained.

FIG. 10: *A Double Spanish Burton (second method).* Five times the power is gained.

FIG. 11: *A Three-Fold Purchase.* Six times the power is gained.

Terminology of Block and Rope Tackle

Bell purchase: Four single blocks.

Block: A wooden or metal frame in which are fitted sheaves or pulleys over which rope is led.

Boom tackle: Used to guy out booms on a fore and aft rigged ship while sailing before the wind.

Breech: End of block opposite swallow.

Burton: Used for heavy lifts on deck or from the mast where it is hooked to a pendant near the head.

Cheeks: Side pieces on the frame of a block.

Chump block: Egg-shaped block with rounded shell.

Deck tackle: A two-fold heavy purchase used in connection with ground tackle, etc.

Double luff: A double and treble block.

Double Spanish Burton: Three single blocks or one double and two single blocks.

Fall: That part of a rope tackle to which power is applied for hoisting.

Fiddle block: Same as a Sister Block except that the falls lead the same way.

Fish block: Lower block of a fish tackle fitted with a fishhook for fishing an old-fashioned anchor.

Gin block: Shell serving as guard to hold rope in the score of the sheave.

Gun tackle: Two single blocks.

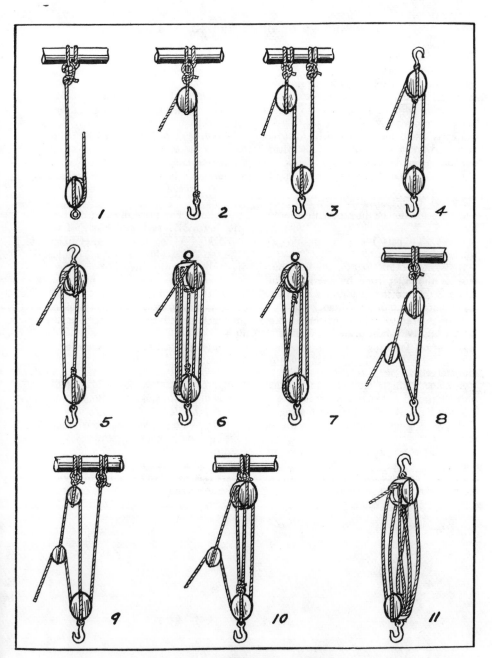

PLATE 46—BLOCKS AND TACKLES

Handybilly: Small light tackle used for miscellaneous work.

Hatch tackle: Used for hoisting and lowering in hatches.

Hauling part: End of falls to which power is applied.

Jigger: Small light tackle used for a variety of work aboard ship, such as lifting sails up the yards, etc.

Luff or watch tackle: A single and double block.

Luff upon luff: A luff tackle used in connection with the hauling part of another luff.

Reeve: To pass the rope around the wheel or sheave of a block.

Relieving tackle: Used for the same purpose as the tiller rope in steering. This tackle is rove on the quadrant when there is too much strain on the wheel.

Rolling tackle: Used on quarter of lower yards to relieve strain in heavy weather.

Round in: To bring two blocks together.

Rudder tackle: Used on rudder pendants for emergency.

Runner: A single movable block which adds additional power to a purchase.

Secret block: Used to prevent fouling.

Sheave: A wheel or pulley over which a rope runs.

Sheet block: Half a shell covering the sheaves.

Single Spanish Burton: Two single blocks and a hook.

Sister block: Two sheaves in one shell fitted for leading falls in opposite directions.

Snatch block: A single sheave block that is used to give a fairlead to the hauling part of a tackle.

Standing part: That part of the fall made fast to a block.

Stay tackle: See *Yard tackle.*

Stock and bill tackle: Used for hauling anchors from and to the billboard in bygone days.

Swallow: The space used for the rope to pass between sheave and frame of block.

Tackle: A combination assemblage of ropes and blocks for the purpose of increasing pull or multiplying force.

Three-fold purchase: Two treble blocks at right angles to each other.

'Thwartship tackle: A term applied to any tackle used for thwartship or across deck.

Topping lift: A rope tackle leading from the masthead of a ship for raising or topping booms.

Two-blocked: Two blocks in close unity.

Two-fold purchase: Two double blocks.

Watch tackle: See *Luff tackle.*

Whip: A single block and rope purchase.

Whip and runner: A whip hooked to hauling part of a runner.

Yard or stay tackle: Used over a hatch where it is made fast to a stay while hoisting and transferring stores.

Terminology

The following names, definitions, and terms apply to the various kinds and types of ropes and knots and their applications:

Rope Construction

Rope is made by twisting fibers into yarns or threads, then twisting the yarns into strands, and finally twisting the strands together to form the finished rope. As the rope is built up each part successively is twisted in an opposite direction; thus, when the yarns are twisted in a right-hand direction, the strands are twisted left-handed and the rope is twisted right-handed. This forms a right-handed plain laid rope. If three or more of these right-handed plain laid ropes are used as strands to form another rope, it will be a left-handed hawser, or cable-laid rope.

When a rope has four or more strands, it is customary to put a core or line in the center to retain the rounded form of its exterior, and this core or line is called the heart.

Rope is designated as right-laid or left-laid, according to the direction in which the strands are twisted. To determine which way the rope is laid, look along the rope, and if the strands advance to the right, or in a clockwise direction, the rope is right-laid; while on the other hand if the strands advance to the left or in a counter-clockwise direction the rope is left-laid.

Small Cordage

Small cordage: Commonly referred to as small stuff, may mean any small cord or line. However, halyards and other similar lines are not usually referred to as small stuff. Cords are generally designated by the number of threads of which they are made, twenty-four thread stuff being the largest. Aside from being known by the number of threads various kinds of small stuff have names of their own. As:

Spun yarn: This is the cheapest and most commonly used for seizing, serving, etc., where neatness is not important. It is laid up loosely, left-handed, in 2, 3 and 4 strands and is tarred.

Marline: Cord of this kind has the same applications as spun yarn but it makes a neater job. It is two-stranded and laid up left-handed. Untarred marline is used for sennit, a braided cord or fabric made from

plaited yarns. Tarred yacht marline is used in rigging lofts.

Houseline and roundline: These lines are used for the same purpose as marline. They are three-strand cord; houseline being laid up left-handed and roundline right-handed.

Hambroline: This is right-handed, three-stranded small stuff made of fine back-handed untarred hemp yarns. Also called hamber.

White line or Cod line: These lines are small stuff made of untarred American hemp or cotton.

Seizing stuff: This is a heavier line than any of the other small stuff and it is used when a strong neat job is required. It is a finished machine-made rope, commonly three-strand, of right-hand stuff. There may be 2, 3 or 4 threads to the strand, making 6, 9 or 12-thread seizing stuff. Tarred American hemp is the material used in its manufacture.

Ratline stuff: This is much the same as seizing stuff, except that it is larger. Its sizes usually run from 6 to 24 threads.

Foxes: This is comprised of two yarns hand-twisted, or one yarn twisted against its lay and rubbed smooth with the hands against the knee.

Rope Materials

Fiber rope: Under this heading are such materials as Manila, hemp, cotton and flax. It takes its name from the species of the plant from which the fiber is taken. Fiber rope is impregnated with oil when manufactured, which adds about ten per cent to its actual weight. The oil adds to the life of the rope by keeping out heat and moisture. As the oil leaves the rope the latter tends to deteriorate rapidly. The strength of fiber rope decreases with usage and a used rope is often deceptive, in that it may not be as serviceable as it looks. Unlike a wire rope the strands of fiber rope do not wear flat, thereby giving a visible sign of weakness. The fibers stretch and twist, but this does not always indicate decreased strength. It is not advisable to place a maximum strain on a rope that has been under a load for any considerable length of time, or one that has been strained to near its breaking point. The safety of rope decreases rapidly with constant use, depending, of course, upon the circumstances and the amount of strain to which it is subjected.

Abaca, or Manila rope: This is the most important cordage fiber of the world today. It possesses a lightness and strength with which no other fiber can compare. Salt or sea water has but little effect upon it and therefore it is used almost exclusively for marine cordage. The material from which it is made is taken from the fibers of the abaca plant and its principal source is Manila, hence its name.

Hemp or sisal: The fibers of these plants are used extensively in rope making. The plants grow abundantly in Italy, Russia, the United States and Mexico. Most of the rope used in the United States Navy is

---

NAMES OF TOOLS SHOWN IN PLATE 47

1: Fid	10: Meshing gage
2: Rope knife	11: Small sail needle
3: Scissors	12: Small meshing gage
4: Thimble	13: Meshing needle
5: Beeswax	14: Large bag needle
6: Round thimble	15: Sail twine
7: Small bag needle	16: Sail hook
8: Palm	17: Meshing needle
9: Sailmaker's needle	18: Small meshing needle

Refer to the Glossary in the following pages for further definitions of these tools.

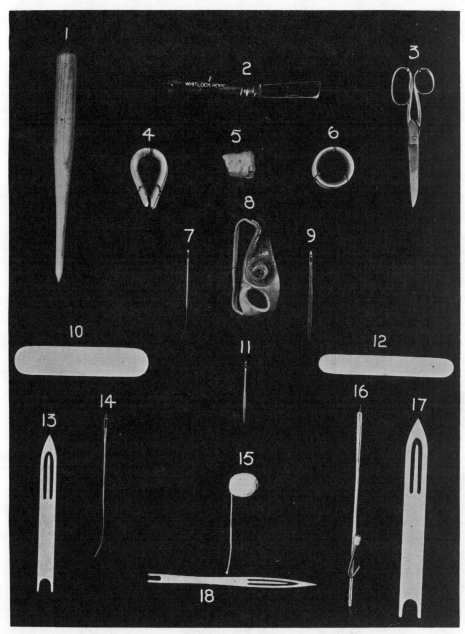

PLATE 47—TOOLS USED IN ROPE AND CANVAS WORK
(*See opposite page for names of tools.*)

made from the American hemp, which is equal in quality to that of any of the other countries from which the fibers are obtained. However, hemp rope is but little used, except for standing rigging, and seldom for that today, since most of the present-day standing rigging is made of wire rope. When used as rigging it is always tarred, the tar tending to protect it from moisture and increase its life, but decreasing its strength.

Coir: Coir is obtained from the fibrous husks of coconuts. Rope made from it is buoyant and does not become waterlogged easily. It has about one-half the strength of Manila rope and finds its chief usage in light lines.

Cotton and flax: These materials are both employed in rope making, although they find but little use aboard ship, except as small stuff. The taffrail log, signal halyards and lead lines are made of cotton, while braided flax is used for boat lead lines.

Notes on the Care and Handling of Rope

To open a coil of rope may seem very simple, yet many a seaman has found himself in "trouble" with a new line, because he did not stop to think before he grabbed an end blindly and started to measure off the amount he wanted. To prevent kinks, first inspect the coil and locate the *inside* end, which is within the eye. (By the expression eye is meant the opening in the center of the coil.) Turn the coil over so that the inside end is down; reach down inside of the eye and take hold of the end; pull it up through the eye and as it comes out it should uncoil counter-clockwise.

Rope manufacturers recommend that the lashings around a coil of rope should be cut from the inside of the eye, while the burlap coverings are left on the out-side of the coil. Rope shrinks in length when it is wet. If it is held taut in dry weather it will be subjected to a great strain when it becomes wet, sometimes so great that it will part. Taut lines should be slacked off when they become wet. Even a heavy dew at night will penetrate an old line, and for this reason running rigging should be eased off at night.

Both heat and moisture will cause rope to lose its strength. To avoid this, rope should always be stored in a cool but dry place. Wet rope should never be stored, nor should it be covered in a manner that will retain the moisture it contains. Rope should be covered to protect it from the weather, and as protection against chafing it should be parceled.

Coiling Rope

Straight coil: Lay a bight of the secured end on deck and lay additional bights on top of it, using up the entire amount of line; keep out all kinks and coils. Turn the entire coil over and it will be clear for running.

To *Flemish* down a line, make a small circle of the free end and continue to lay small circles around it until the entire line is down and has the appearance of a coiled clock spring.

To *fake* down a line, lay the free end out in a straight line, then turn back a loop to form a close flat coil; continue to lay flat coils with the ends on top of the ends of the preceding coil. Always coil a line with the lay.

Right-handed rope should always be coiled "with the sun," that is, in a clockwise direction.

Left-handed rope should always be coiled "against the sun," that is, in a counter-clockwise direction.

Glossary

In its true sense the word "knot" is any fastening made by interweaving rope or cordage, yet in the common sense the term is generic, in that it is considered as including many kinds and varieties of actual knots, as well as other rope fastenings, such as are known as bends, hitches, and splices.

The three principal classes of knots, in the commonly accepted meaning of the term, are:

Bend, from the same root-word as "bind," is defined as a method of joining the ends of two ropes together, or in the language of the sea, the bending of two ropes together, as the Sheet Bend, Carrick Bend, etc.

Hitch, which is an Old English word, is defined as a method of securing a rope to some object such as a spar, as the Clove Hitch, or to another rope, as the Rolling Hitch.

Knot, from the Anglo-Saxon *cnotta,* re-fers to a method of forming a knob in a rope, by turning the rope on itself through a loop, as in the case of the Overhand Knot, the Bowline Knot, and the Running Knot.

The use of these three terms in connection with the word "knot" appears to be arbitrary, since there is but very little difference between the Fisherman's Bend and the Timber Hitch. In one sense the terms knot and seizing are considered as permanent fastenings, which must be unwoven or disentangled in order to be unfastened, while the bend or hitch can be unfastened by merely pulling the ropes in the opposite or reverse direction to that in which they are intended to hold.

There are many other terms used in connection with the tying of knots and with the uses of rope and cordage, most of which are given in the following glossary. This is quite general in its scope and contains many terms which do not apply to knots and rope, but which are closely related to their applications.

Knot and Rope Terms

Abaca: A plant which grows chiefly in the Philippine Islands, from which the fibers are taken for making so-called Manila rope.

Abaca rope: Rope made from the fibers of the abaca plant; also called Manila rope, since most of the abaca fibers are obtained from that source.

Against the sun: A nautical term meaning rotation in a counter-clockwise, or left-handed direction, in contradistinction to clockwise, or right-handed, as in the direction of rotation of the sun.

Back-handed rope: In rope making the general practice is to spin the yarn over from right to left, or counter-clockwise, making the rope yarn right-handed. The strand formed by a combination of such yarns becomes left-handed and three of these strands twisted together form a right-handed rope, or a plain-laid rope.

In making back-handed rope if, instead of twisting the strand in a direction opposite to the direction of the twist of the yarn, it is twisted in the same direction, that is, right-handed, then when brought together and laid up the rope is left-handed. This is called left-handed or back-handed rope. It is considered as being

101

more pliable than plain-laid rope and less liable to kinks.

Baseball stitch: A form of stitching or sewing used in sail making and repairing.

Becket: A rope eye for the hook of a block. Also; a rope grommet used as a rowlock or any small rope strop used as a handle.

Becket bend: An efficient bend used for uniting the two ends of a rope or the end of a rope to an eye. It jams tight with the strain, will not slip and is easily cast off.

Beeswax: This wax matter derived from the honey-comb found in bee-hives is used extensively in canvas work. It is applied to the sail twine to prevent the small threads from fraying as the twine is pulled through the canvas. Usually it is in the form of a wad as in FIG. 5.

Belay: To make a rope or line fast by winding it in figure-eight fashion around a cleat, a belaying pin, or a pair of bitts. Also, to stop or cease.

Belaying pin: A pin of either wood or metal set in such places as pin rails, etc., upon which to belay a rope or secure the running rigging.

Bend: As defined previously. Also, to secure, tie or make fast, as to bend two ropes together.

Bight: A loop in a rope, as that part of the rope between the end and the standing part, formed by bringing the end of the rope around, near to, or across its own part.

Bitt: One or more heavy pieces of wood or metal, usually set vertically in the deck of a vessel, for the purpose of securing mooring lines and tow lines.

Bitt a cable: To make a line fast by a turn under the thwartship piece and again around the bitt-head; or to double or weather-bitt a cable an extra turn is taken.

Bitt head: The upper part of a bitt, or its head or top.

Bitter-end: The last part of a cable that is doing useful work. In the case of an anchor chain its bitter-end is made fast in the bottom or side of the chain locker.

Block: A mechanical device consisting essentially of a frame or shell, within which is mounted a sheave or roller over which a rope is run. There are many varieties of blocks which are at times called pulleys, or when rigged, a block and tackle. The name pulley, as used in connection with a block, is a misnomer in that in this case the word pulley refers only to the sheave or roller.

Bollard: A heavy piece of wood or metal set in the deck of a vessel or on the dock to which the mooring lines are made fast. They are also called nigger heads.

Bolt rope: A piece of rope sewed into the edge of a piece of canvas or a sail to give added strength and to prevent the canvas from ripping. They are made of hemp or cotton cordage and the name is now applied to a good quality of long-fiber, Manila or hemp.

Braid: To plat, plait or interweave strands, yarns, ropes or cords.

Breast line: A line leading from the breast of a ship to a cleat on the dock, without leading forward or aft, for the purpose of mooring the ship.

Bridle: Any span of rope with its ends secured.

Cable: A heavy rope used in attaching anchors or in towing. A cable is also a nautical measure of length. (*See* Cable-length.)

Cable-laid: Cable-laid rope is made up of three ropes laid up left-handed; the ropes comprising the strands being laid up right-handed. It is also called hawser-laid rope.

Cable-length: A nautical measure of length. Its name was derived originally from the length of a ship's cable which bears no relation today to the length of any cable. Many recognized authorities on the subject differ widely in their explanations of the definition of the term. Most of them, however, define a cable-length as being equal to 120 fathoms, although in some instances it is given as 100 fathoms. Admiral Luce, in his work on *Seamanship*, published in 1863, gives the most enlightening definition of a cable-length with a plausible reason for it. His explanation is that custom limited the length of cables to 120 fathoms for the reason that rope walks of earlier times were unable to lay up strands of greater length. This was due to the length of the early rope walks. To lay up a cable longer than 120 fathoms would require that the rope walks be one-third longer than the cable, as this extra length would be needed for drawing down the yarns and laying up the strands. Other difficulties in rope making contributed too as factors in limiting the lengths of rope walks and hence cable lengths also.

Canvas: A woven fabric of cotton or flax used for sails, awnings and many other shipboard purposes. It is made in varying degrees of weight or quality, numbered from 00, the coarsest, to 10, the finest weave. The term in nautical parlance is synonymous with sail.

Catch a turn: To take a turn, as around a capstan or bitt, usually for holding temporarily.

Caulking: To drive oakum or cotton into the seams of the deck or the ship's side, as a means for preventing leakage. Also, the pieces of oakum or cotton used for caulking purposes.

Cavil: A strong timber, bollard, or cleat used for making fast the heavier lines of a vessel. Also called kevel.

Chafe: To rub or abrade.

Chafing gear: A winding of small stuff, rope, canvas, or other materials around spars, rigging, ropes, etc., to prevent chafing.

Chafing mat: A mat made from woven ropes or cordage to prevent chafing.

Clap on: A nautical term with several meanings, as: to seize or take hold of; clap on more sail, and clap on a Wall or Crown on the end of a rope.

Cleat: A heavy piece of wood or metal having two horns around which ropes may be made fast or belayed. Usually secured by bolts or lashings to some fixed object, as the deck of a vessel or the dock.

Clinch: A form of bend by which a bight or eye is made by seizing the end to its own part. Two kinds of clinches are used, an inside and an outside clinch. Also, an oval washer, at times called a clinch ring, which is used on spikes and bolts where they are employed in wooden construction.

Clockwise: Rotation in the direction of the hands of a clock, or right-handed, in contradistinction to counter-clockwise or left-handed.

Clothes stop: Small cotton line used for stopping clothes to a line or for securing clothes rolled up in bags or lockers.

Cod line: Small stuff made of eighteen thread untarred American hemp or cotton.

Coil: A series of rings, or a spiral, as of rope, cable and the like. Most rope is sold by the coil, which contains 200 fathoms. Also, to lay a rope or cable down in circular turns. If a rope is coiled right-handed, that is, in the direction of rotation of the hands of a clock, it is coiled from left to right, or with the sun. Hemp rope is always coiled in this manner.

Coir: The fibers of the outer husks of the coconut.

Coir rope: Rope made from coir fibers. It is extremely light in weight but is not as strong as rope or cable made from the other common rope materials.

Concluding line: A small rope or line rove through the middle of the steps of a Jacob's ladder.

Cord: Comprised of several yarns, usually cotton or hemp, with an extra twist, laid up the opposite way. Also a term employed as a distinction between small stuff and rope.

Cordage: A general term now commonly employed to include all ropes and cords, but, more specifically, it applies to the smaller cords and lines. Also used as a collective term in speaking of that part of a ship's rigging made up of ropes, etc.

Core: A small rope run through the center of heavier rope. It is usually found in four-strand rope, lending to it a smooth, round outside appearance.

Cotton: The fibers of the cotton plant, used in making ropes and small stuff.

Cotton canvas: Fabric made from cotton yarns.

Cotton rope: Rope made from cotton fibers or yarns.

Counter-clockwise: Opposite in rotation to the hands of a clock, or left-handed, in contradistinction to right-handed or clockwise.

Cow's tail: Frayed end of rope. Same as dead men and Irish pennant.

Creasing stick: A wood or metal tool slotted at one end, used for creasing or flattening seams in canvas preparatory to stitching.

Cringle: A piece of rope spliced into an eye over a thimble in the bolt rope of a sail.

Cross turns: Turns taken around a rope at right angles to the turns of the lashings or seizing.

Crown: In knotting to so tuck the strands of a rope's end as to lock them in such a manner as to prevent unraveling by back-splicing the strands. A Crown over a

Wall Knot is the first step in tying a Man-rope Knot.

Dead men: The frayed ends of ropes. Same as cow's tail and Irish pennant.

Dead rope: A rope in a tackle not led through a block or sheave.

Earing: A short piece of rope secured to a cringle for the purpose of hauling out the cringle.

Elliott eye: A thimble spliced in the end of a cable or hawser.

End: In knotting that part of a rope extending from the bight to its extremity, in which the standing part of the rope is on one side of the knot and the end on the other.

End seizing: A round seizing on the ends of ropes.

Eye: A loop in the end of a rope, usually made permanent by splicing or seizing.

Eye seizing: A seizing used for shortening an eye, the turns of the seizing stuff being taken over and under each part and the ends crossed over the turns.

Fag end: An unraveled or untwisted end of rope, or as applied to flags and pennants, the ragged end.

Fair leader: Lines employed for the purpose of leading other lines in a desired direction. That is, they may be used to pull a line at an angle for the purpose of securing it to a belaying pin or cleat.

Fake: A circle or coil of rope in which the coils overlap and the rope is free for running. Also to fake down a rope is to coil down a rope.

Fall: A rope, which with the blocks makes up a tackle. A fall has both a hauling part and a standing part, the latter being the end secured to the tail of the block. In some cases only the hauling part is considered as the fall.

Fast: Attached or secured to. Also to make fast means to secure or attach.

Fiber: The smallest thread-like tissue of a rope, cord, or thread, as the fibers of flax, cotton, hemp, Manila, etc., which are twisted to make the yarn out of which the rope is made.

Fiber rope: Rope made from the fibers of abaca, hemp, coir, Manila, sisal, etc.

Fid: A tapered piece of hardwood or a pin similar in form to a marline spike and used for the same purpose, such as separating the strands of a rope in making splices (Fig. 1).

Flat seam: The most common method of sewing two pieces of canvas together. The seam is made by lapping the two pieces of canvas and then pushing the needle through the lower and up through the piece to be joined.

Flax: The fibers of the flax plant twisted into yarns for the purpose of making ropes, cords, and small stuff, as well as canvas fabric.

Flax canvas: Fabric made from yarns twisted from flax fibers.

Flax rope: Rope made from yarns twisted from flax fibers.

Flemish coil: There are several versions of the exact meaning of the term. One is that it is a coil in which each fake rides directly on top of the fake below it. Another version is that a Flemish coil, also called a Flemish fake or mat, is one in which each fake lies flat and concentrically, giving the appearance of a rope mat. Common practice, however, is to make a concentric rope mat when ordered to Flemish down a coil and when the order is to fake, or fake down, it is the practice to lay down a coil with one fake upon the other, for clear running.

Flemish eye: An eye in which the strands are not interwoven but are separated and bound down securely to the main part with seizing.

Foot rope: A stirrup or strap hanging in a bight beneath a yard in which the men stand when furling or reefing sail.

Foxes: As previously explained, two yarns hand-twisted, or one yarn twisted against its lay and rubbed smooth.

Frap: To bind or draw together and secure with ropes, as two slack lines.

Frapping turns: Same as cross turns.

Freshen the nip: To shift or secure a rope in another part to minimize wear.

Gantline: A whip or light line made fast aloft for the purpose of hoisting boatswain's chairs and for hoisting sails aloft for bending. Also called girtline.

Garland: A strop used primarily for the purpose of hoisting spars.

Garnet: An arrangement of tackle used for hoisting cargo.

Gasket: The term has several nautical meanings. In one sense a gasket is a rope or band of canvas by means of which sails are made fast to the booms or yards. Canvas and sennit bands used to secure a sail are called harbor gaskets, while a rope wound around a sail and a yard arm to the bunt is called a sea gasket. Gaskets are also defined as any kind of packing employed for making joints water- or gas-tight.

Geswarp: A hauling line laid out by a boat, a portion of which is coiled down in the boat. Also called guess warp, or a line for mooring a boat to a boom.

Girtline: Same as gantline.

Gooseneck: A bight made in the standing part of a rope in forming a bowline.

Grab rope: A rope used as a handrail, such as a line secured waist high above a boat-boom or gang-plank for steadying one's self.

Grafting: To cover a strap, ring-bolt or

any similar article with log-line, cord or small stuff.

Graft a rope: To taper the end of a rope and cover it completely with an ornamental arrangement of Half Hitches of small stuff.

Grommet: Eyelets made of rope, leather, metal, and other materials. Their chief uses are as eyelets secured to canvas and sails, through which stops or robands are passed.

Guess warp: Same as Geswarp.

Guest rope: A grab rope running alongside a vessel to assist boats coming alongside.

Guy: Any rope used for steadying purposes.

Gypsy: The drum of a windlass or winch around which a line is taken for hauling in.

Halyard: Any of the small ropes and tackles used for hoisting sails, flags and the like.

Hamber: Three-stranded seizing stuff tight-laid, right-handed. Also called Hambroline. It is usually made of untarred hemp and is similar to roundline.

Hambroline: Same as hamber.

Hand rope: Same as grab rope.

Hauling line: Usually a line sent down from aloft for the purpose of hauling up gear. Also: any line used for hauling purposes.

Hauling part: That part of the rope in a tackle which is hauled upon, or it might be described as the end of the falls or a rope to which power is applied.

Hawser: Any large rope, five or more inches in circumference, used principally for kedging, warping and towing.

Hawser-laid: Left-handed rope of nine strands laid up in the form of three, three-stranded, right-handed ropes.

Heart: Same as core

Heave: To haul or pull on a line or to cast a heaving line.

Heave taut: To pull tight or stretch.

Heaving line: A light flexible line having a monkey fist on its end. It is heaved ashore or to another vessel as a so-called messenger line, as when it is desired to pass a mooring line to a dock.

Hemp: The fibers of the hemp plant used in making rope. The fibers of this plant are obtained from such sources as Italy, Russia, America and other countries. That obtained from New Zealand is called Phormium hemp, and that from the East Indies is known by the name of Sunn hemp. Rope made of hemp fibers is usually tarred, but if not it is called white rope. American hemp is distinguished by its dark gray color.

Hemp rope: Rope made from the fibers of the hemp plant.

Herringbone stitch: A form of stitching or sewing used in sail-making and in the sewing of canvas.

Hide rope: Rope made from strips of leather cut from green hides. It finds but little use today, except as wheel rope.

Hitch: As previously explained, or a combination of turns for securing a rope to a spar or stay.

Hook rope: A rope with a hook secured to its end used for clearing cable chains, etc.

House line: Three-stranded cord or small stuff laid up left-handed and used for seizings, etc. Sometimes called round line, although the latter is laid up right-handed.

Irish pennant: Same as cow's tail and dead men.

Jam: To wedge tight.

Junk: Salvaged rope which is made up into swabs, spun yarn, nettle-stuff, lacings, seizings, earings, etc., although little used today.

Jute: The fibers obtained from East Indian plants used in making sacking, burlap and twine.

Jute rope: Rope made from jute fibers.

Keckling: Chafing gear applied to a cable, usually made up of old rope.

Kedging: Moving a vessel by alternately laying out a small anchor or kedge and hauling the vessel up to it.

Kevel: Same as cavil.

Kink: A twist in a rope.

Knife: Knives form an important part in the working of rope and canvas. One such as is used commonly by sailors is shown in FIG. 2.

Knittles: Rope yarns twisted and rubbed smooth for pointing and similar purposes. Two or more yarns may be twisted together or one yarn may be split and its halves twisted. Also called nettles.

Knot: As previously described, refers to a method of forming a knot in a rope, by turning the rope on itself through a loop.

Lacing: Small rope or cord used to secure canvas or sails by passing it through eyelets in the canvas.

Lang-lay rope: Wire rope in which the individual wires comprising the rope are twisted in the same direction as the strands which make up the rope.

Lanyard: Four-strand tarred hemp rope used for making anything fast. Also an ornamental braid or plait used by sailors in securing their knives and the like.

Lash: To secure by binding with rope or small stuff.

Lashing: A binding or wrapping of small stuff used to secure one object or line to another, as an eye to a spar.

Lashing eye: Loops in the ends of two ropes through which are passed the lashings which bind them together.

Lasket: Loops of small cord used to lace the bonnet to the jib. Sometimes called latchings.

Latching: Same as lasket.

Lay: The direction in which the strands of a rope are twisted. This may be either right-handed (clockwise) or left-handed (counter-clockwise). It also refers to the degree of tightness with which the strands are twisted, as soft, medium, common, plain and hard lay. Also used in the expressions "against the lay" and "with the lay" as denoting a direction contrary to or with the lay of the strands of the rope.

Leading part: Same as hauling part.

Lead line: A line secured to the ship's lead which is used for sounding the depth of water under a vessel. A lead line is made of braided cotton twine for a boat; braided flax for a ship, and of hemp for deep sea purposes. Now being replaced with fine wire, wound on reels. (Pronounced led.)

Left-handed: Direction of rotation from right to left, counter-clockwise or against the sun, in contradistinction to right-handed, clockwise or with the sun.

Left-handed rope: Same as back-handed rope.

Line: Any rope or line, the word being more commonly used in marine practice than the term rope.

Log line: A line attached to a ship's log, an indicator for measuring speed and distance. The taffrail log line is made of braided cotton twine while the chip log line is made of hemp.

Loop: Same as bight.

Man rope: A rope hung over the ship's side for the purpose of ascending or descending.

Manila: A term used to describe rope made from the abaca fiber, which is obtained chiefly from Manila.

Manila rope: Rope made from the fibers

of the abaca; the chief source is Manila or the Philippine Islands.

Marl: To make secure or bind with a series of Marline Hitches.

Marline: See definition under Small Cordage.

Marline spike: A metal pin tapering to a point used chiefly for splicing wire rope, etc.

Marry: Binding two lines together temporarily, either side by side or end to end.

Match rope: An inflammable rope used as a fuse.

Meshing gage: These gages are used in making the meshes in nets. They may be of any length to suit the user's convenience, but their width determines the size of the mesh in the net. Thus, a meshing gage one inch wide will form a mesh one-half-inch square. (*See* Figs. 10 and 12.)

Meshing needle: Needles such as these are usually made of wood, but many have been made from light metal. They are used in the form of a shuttle for passing the cord and tying knots in the making of nets. In winding the cord on the needle it should be understood that both the needle and the cord it contains must be small enough to pass through the mesh of the net. They may be made of any desired length. Several forms are shown in Figs. 13, 17, and 18.

Messenger line: A light line such as a heaving line used for hauling over heavy lines, as from a ship to a dock.

Middle a rope: Doubling a rope in such a manner that the two parts are of equal length.

Mooring line: A line such as a cable or hawser used for mooring a vessel.

Mousing: A short piece of small stuff seized across the opening of the hook of a block as a measure of safety.

Nettles: Same as knittles.

Nip: A short turn in a rope or to pinch or close in upon.

Painter: A short piece of rope secured to the bow of a small boat used for making the boat fast. Not to be confused with sea painter.

Palm: A sailor's thimble, so to speak. It is comprised of a leather strap worn over the hand to which is attached a metal plate so placed as to fit into the palm of the sailor's hand. It is used primarily in sewing and mending sails and canvas (Fig. 8).

Parbuckle: A form of sling consisting of two ropes passing down a ship's side for the purpose of hauling a cask or similar object in the bight of the ropes, one end of each rope being secured and the other tended.

Parcel: To protect a rope from the weather by winding strips of canvas or other material around it with the lay preparatory to serving.

Part: To break.

Pass a lashing: Make the necessary turns to secure a line or an object.

Pass a line: Carry a line to or around something or reeve through and make fast.

Pass a stopper: To reeve and secure a stopper.

Payed: Painted, tarred or greased to exclude moisture, as payed rope.

Pay out: To slack off on a line or let it run out.

Pendant: A length of rope with a block or thimble secured to its end.

Plain laid: Rope in which three strands of left-handed yarn are twisted together to form a right-handed or plain laid rope.

Plait: Same as braid and plat.

Plat: Same as braid and plait.

Pointing: Any of numerous methods by which the end of a rope is worked into a stiff cone-shaped point.

Pricker: A light piece of metal, similar to a marline spike, but having a handle and used for the same purposes as a marline spike.

Quilting: Any covering of woven ropes or Sennit placed on the outside of a container used for water.

Rack: To seize two ropes together with cross turns of spun yarn or small stuff.

Rack seizing: Small stuff rove around two lines in an over and under figure-eight fashion.

Ratline: As previously described, three-strand, right-hand, tarred small stuff larger than seizing stuff. Also short lines of ratline stuff run horizontally across shrouds as steps.

Reef: To take in sail to reduce the effective area of a sail.

Reef cringle: A rope grommet worked into a thimble attached to a sail.

Reef earing: A short length of small stuff which is spliced into a reef cringle for the purpose of lashing the cringle to a boom or yard.

Reef points: Short pieces of small stuff set in the reef bands of sails for reefing purposes.

Reeve: To pass the end of a rope through an eye or an opening, as through a block, thimble, or bight.

Render: To pass through freely. Said of a rope when it runs easily through a fairlead or a sheave.

Riders: A second layer of turns placed over the first layer of turns of a seizing.

Riding chock: Chocks over which the anchor chains pass.

Riding turns: Same as riders.

Rigging: A term applied to a ship's ropes generally.

Right-handed: Direction of rotation from left to right, as clockwise and with the sun, in contradistinction to left-handed, counter-clockwise or against the sun.

Right-handed rope: Same as plain laid rope.

Robands: Short pieces of small stuff of Manila or spun yarn, used to secure the luff of a sail. Also called rope bands.

Rogue's yarn: Colored yarn worked into the strands of a rope as an identification mark and as protection against theft.

Rope: A general term used to describe any cord more than one inch in circumference. In this connection all rope was at one time measured by its circumference, except that used for power transmission, and it was measured by its diameter. Today, however, all rope is usually measured in terms of its diameter.

Rope band: Same as roband.

Roping needle: A short spur needle used for the purpose of sewing bolt ropes and other heavy work.

Roping twine: Nine- to eleven-ply small stuff.

Roping yarn: Untwisted strands of a rope used for rough seizings.

Rounding: Serving a rope to prevent chafing.

Round line: Three-stranded cord or small stuff laid up right-handed. Also called house line.

Round seam: A seam employed in sewing canvas and sails. It is used to join two edges together by holding them together and passing the needle through them at right-angles to the canvas.

Round seizing: A method of securing two ropes together or the parts of the same rope to form an eye.

Round turn: To pass a line completely around a spar, bitt or another rope.

Rouse: Heave heavily on a line.

Rubber: A flat steel tool used for rubbing or smoothing down seams in canvas. It is fitted with a wooden handle.

Rumbowline: Coarse soft rope made from outside fibers and yarns and used for temporary lashings, etc.

Running line: Any line such as a messenger line, that is, such a line as might be run out for paying out a hawser or cable.

Running rigging: All the lines of a ship used to control the sails.

Sail: Generally, canvas used aboard ship in the form of sails.

Sail hook: A metal hook secured to a small line, used for holding sails and canvas while sewing seams, etc. (*See* Fig. 16.)

Sail needle: Long spur steel needles, triangular at the point and cylindrical in form at the eye. There are two types, known respectively as sewing and roping needles, and these are in turn called short and long spur needles. Sewing needles come in the sizes 6 to 14 and in half-sizes to $17\frac{1}{2}$. Size 15, which is $2\frac{1}{2}$ inches long, is the most used. (Fig. 7 is a small bag or tufting needle; Fig. 9, a sailmaker's needle; Fig. 11, one of the smaller needles, and Fig. 14 is a large bag or tufting needle.)

Sail twine: Small stuff in the form of thread made of either cotton or flax. It is available in several different sizes, usually wound into the form of a ball (Fig. 15).

Scissors: These are essential in canvas and fancy rope work. A pair is shown in Fig. 3.

Sea painter: A long rope, not less than $2\frac{3}{4}$ inches in circumference, for use in a ship's lifeboats.

Secure: To attach, fasten or make fast.

Secured end: That part of a line, or the end of a line which is attached or secured to an object.

Seize: To put on or clap on a seizing. That is, bind with small stuff, as one rope to another, a rope to a spar, etc. Seizings are named from their appearance or from the functions they serve, as throat seizing, flat, and round seizing, and others.

Seizing stuff: Small stuff.

Selvage: A woven border in fabric, or the edge of the fabric so closed by complicating the threads as to prevent raveling.

Selvagee: Rope yarn, spun yarn or small stuff marled together and used for stoppers, straps, etc.

Sennit: Braided cordage such as nettles or small stuff, usually of ornamental design, the different kinds being known by many different names, according to their design.

Serving: To wind a rope or wire, after worming and parceling with small stuff, the turns being kept very close together to make the end of the rope impervious to water after being tarred. The turns of the serving are made against the lay of the rope. Also called service.

Serving board: A small flat board having a handle, used with a serving mallet for serving rope. The mallet is a typical round, wooden maul with a groove cut lengthwise in its head. Their purpose is to keep the serving taut and the turns close together.

Shakings: Odds and ends of rope and scrap rope which are hand-picked into fibers for use as oakum.

Sheave: The roller of a tackle block.

Shoemaker's stitch: A sewing stitch used in making and mending sails and in sewing canvas and other fabrics.

Shroud-laid: Four-stranded rope laid up right-handed.

Shrouds: Ropes of hemp or wire used as side stays from the masthead to the rail.

Sisal: Fibers of the henequin plant which grows abundantly in Yucatan.

Sisal rope: Rope made from fibers of the sisal plant.

Slack: That part of a rope between

its secured ends which is hanging loose; the opposite of taut.

Sling: To sling a piece of cargo, as a barrel, by passing a line around it. Or, a sling is an arrangement of short pieces of rope or chain used for hoisting cargo and other similar purposes. Of these the rope sling or strap is the most common. A rope sling is made of short pieces of rope spliced together. It is passed around a cask, for instance, and one bight, the rove, is passed through the other, which is known as the bite. The cargo hook is attached in the rove. Other types of slings are known by such names as web, chain, platform, net, bale, butt, etc.

Small Cordage: As previously described; small stuff which may be 1¾ inches in circumference or less.

Spanish fox: Untwisted rope yarn retwisted in the opposite direction.

Spar: A yard, mast, gaff, or boom.

Splice: To join the ends of ropes together, or the end of a rope to its standing part, as in forming an eye, by interweaving the strands of the rope into themselves. Splices are known as long, short, chain, sailmaker's, etc.

Spun yarn: Two- three- and four-stranded rough stuff twisted loosely together and laid up left-handed for use as small stuff, especially in applications where a smooth service is desired.

Standing part: In knotting, the standing part of a line is that part of the main rope as distinguished from the bight and the end.

Standing rigging: That part of a ship's rigging used to support its spars which is not altered with the ordinary working of the ship.

Stay: A piece of rigging, usually a rope of hemp or wire used as a support for a mast.

Stop: To seize or lash, usually temporarily.

Stopper: A short line, one end of which is secured to some fixed object, used to check or stop a running line.

Strand: Part of a rope, made up of yarns, the twist of the strand being opposite to that of the yarns.

Stranded: A rope is said to be stranded when a strand parts.

Strap: A form of sling, usually employed for handling cargo. Also called a strop. Permanent straps are spliced into the thimble with the hook of a tackle block. A sail strap is one in which the eye is made by a seizing and the bight is tucked through the eye instead of the strap itself.

Strop: Same as strap.

Tackle: An arrangement of ropes and blocks, sometimes called block and tackle. Also pronounced tay-kle. Tackle is employed primarily for hoisting purposes.

Tackline: A short length of line, usually signal halyard, used for separating strings of signals.

Take a turn: To pass a line around a cleat or a belaying pin.

Tapered rope: Rope used in places where most of the strain is taken by one end only. The part which bears the strain is full-sized while the remainder of the line tapers off to the hauling part, which is usually light and pliable.

Tarred fittings: Small stuff.

Taut: Tight; the opposite of loose or slack.

Tend: Man or attend.

Thimble: A grooved piece of metal, either circular or heart-shaped, to receive the eye of a rope (Figs. 4 and 6).

Thread: Yarn or small stuff made into strands. Small rope or cord is often designated by the number of threads or yarns it contains, as 6-thread seizing stuff.

Throat seizing: A seizing used to lash an eye in a rope; to hold it around a thimble, or to seize two parts of rope together at the point where they cross each other.

Thrums: Short pieces of small stuff secured by their bights to pieces of canvas for use as chafing gear. The verb thrum means to attach the small stuff.

Thrum mats: Small pieces of canvas with short pieces of rope sewed to them by a method called thrumming. They are used between the rowlocks and the oars to prevent noise.

Tie: That part of a halyard which hoists a yard. Also to tie a knot.

Tiller rope: Wire rope made of copper, bronze and galvanized wire, laid up left-handed.

Timenoguy: A short piece of rope with a bull's-eye spliced in its end, employed as an intermediate support or guy for stays.

Toggle: A small wooden pin made of hardwood which is inserted into a knot to make it more secure or to make it more readily and quickly unfastened.

Tow line: A cable or hawser used for towing.

Trailing line: Small lines attached to the gunwales of small boats and around the loom of the oars to prevent them from falling overboard when the oars are trailed.

Trice: To haul up or pull taut.

Tricing line: Any small line used for suspending articles.

Turn: A single winding of rope, as around a bollard or cleat.

Twice-laid rope: Rope made from second-hand rope yarns or strands.

Unbend: Cast off or adrift, let loose or free; to untie.

Unlay: To separate the strands of a rope.

Unreeve: The opposite of reeve.

Veer: To permit a rope or chain to run out or slack off.

Veer and haul: To slack off and pull alternately.

Warp: Maneuver a ship as at a dock.

Wheel rope: Rope connecting the steering wheel with the drum of the steering gear. Composed of six strands and a heart, making seven strands in all.

Whip: To lash the end of a rope to prevent it from fraying. Also used to designate different kinds and arrangements of blocks and tackles.

Whipping: The lashings on the end of a rope or the small stuff used for whipping.

Whip upon whip: One lashing over the top of the other.

White line: Small stuff made of untarred hemp or cotton.

With the sun: Rotation in the direction of that of the sun, clockwise, or right-handed, in contradistinction to against the sun, counter-clockwise or left-handed.

Worming: Filling up the lays of a rope with spiral windings of small stuff preparatory to parceling in order to make a round, smooth surface.

Yarn: Any number of threads or fibers twisted together from the various fibers from which rope and cords are made.

Yoke lanyard: A small line rove through or secured to the ends of a yoke.

Index

WIRE ROPE